Phototransferred Thermoluminescence

Online at: https://doi.org/10.1088/978-0-7503-3831-8

Phototransferred Thermoluminescence

Makaiko L Chithambo
Department of Physics and Electronics, Rhodes University,
Grahamstown 6140, South Africa

IOP Publishing, Bristol, UK

ISBN 978-0-7503-3831-8 (ebook)
ISBN 978-0-7503-3829-5 (print)
ISBN 978-0-7503-3832-5 (myPrint)
ISBN 978-0-7503-3830-1 (mobi)

DOI 10.1088/978-0-7503-3831-8

Version: 20240301

IOP ebooks

British Library Cataloguing-in-Publication Data: A catalogue record for this book is available from the British Library.

Published by IOP Publishing, wholly owned by The Institute of Physics, London

IOP Publishing, No.2 The Distillery, Glassfields, Avon Street, Bristol, BS2 0GR, UK

US Office: IOP Publishing, Inc., 190 North Independence Mall West, Suite 601, Philadelphia, PA 19106, USA

To Bertha. A Nachanza.

Contents

Preface

If thermoluminescence (TL) is the proverbial nut, then it is fair to say it has been hammered to pieces. Surprisingly, phototransferred thermoluminescence (PTTL) has been spared the same over-exuberant treatment. Thermoluminescence is a method for studying point defects in insulators and appears under controlled heating of an irradiated material. Changes in electron trapping defects and luminescence sites can readily be monitored in temperature-wavelength-intensity isometric plots. If the thermoluminescence (TL) that ensues from one electron trap is a result of electrons optically transferred from another more stable one, that thermoluminescence is, eponymously 'phototransferred'. Although this is a useful tutorial definition of the process, measurements tailored to produce PTTL in this way are not always successful and there lies the motivation for further studies of the process.

This book is devoted to contemporary developments in phototransferred thermoluminescence and is suitable for researchers with an interest in the study of point defects using luminescence methods. The book opens with a look at the principles of thermal stimulation in order to provide a context for PTTL. The subsequent chapters describe experimental techniques for its measurement, cover analytical methods with emphasis on the treatment of PTTL time-response profiles, and provide a snapshot of studies of PTTL in synthetic, natural and other materials of various research interest. The first five chapters refer to crystalline materials whose stand out feature in our context is that the electron trapping defects are associated with well-defined and discrete energies. The final chapter describes selected examples of materials with quasi-continuous distribution of electron trapping energy levels and also provides pointers, for interested readers, on applications of PTTL.

Advances in measurement, analysis and applications of PTTL need not be circumscribed by corresponding developments in TL. We hope that we have covered sufficient ground to provide impetus for further exploration that would improve understanding of phototransferred thermoluminescence.

Acknowledgements

I am delighted to thank everybody who assisted in seeing this project to completion. Caroline Mitchell, until recently the Senior Commissioning Editor; and Isabelle Defillion, the editorial assistant at the Institute of Physics (UK), ably saw to the progress of this book at every stage of its preparation with patience and impeccable efficiency. Thank you for your encouragement and support. It is a pleasure to thank my collaborator and colleague, Dr Jitumani Kalita, who read and offered his thoughts on the opening chapters despite his own busy schedule.

This book is a concise treatment of phototransferred thermoluminescence (PTTL). Although thermoluminescence itself has been richly and systematically studied, the attention on PTTL has been sporadic. Thermoluminescence is a method for studying point defects in insulators and ensues under controlled heating of an irradiated material. Phototransfer seeks to achieve the same effect by using light of suitable wavelength as a restorative agent. The book opens with a look at the principles of thermal stimulation in order to provide a context for PTTL. This is followed by a description of experimental techniques for its measurement and a discussion of analytical methods with much emphasis on the treatment of PTTL time-response profiles. The last three chapters, which can be read independently of each other, provide a snapshot of studies of PTTL in synthetic materials, natural ones and other materials of interest. The last chapter briefly notes, for interested readers, applications of PTTL.

I am grateful to Rhodes University and The National Research Foundation of South Africa for financial support during the research of some of the work included in the book.

These acknowledgements would be incomplete without a word of gratitude to my wife, Bertha. Thank you for your presence, patience, encouragement and support as I worked on this book.

Author biography

Makaiko L Chithambo

Makaiko Chithambo is a professor of physics and head of the physics department at Rhodes University in South Africa. He studied for his BSc at the University of Malawi, for his MPhil at the University of Sussex and for his PhD at the University of Edinburgh. He is a member of the Academy of Science of South Africa. Until recently, he served as the president of the South African Institute of Physics. His research interests are in the study of point-defects in materials using luminescence methods, with special focus on time-resolved optical stimulation techniques. He has, in addition, long standing interests in thermoluminescence, thermoluminescence spectra and radioluminescence. He is the author of the book *An Introduction to Time-resolved Optically Stimulated Luminescence* (2018). His awards include the Vice Chancellor's Distinguished Research Award and the Vice Chancellor's Distinguished Senior Research Award at Rhodes University.

IOP Publishing

Phototransferred Thermoluminescence

Makaiko L Chithambo

Chapter 1

Introduction

Phototransferred thermoluminescence as jargon has two terms. This chapter will mostly address the second term to establish its theoretical principles in order to properly relate it to 'phototransfer'. The rest of the book dwells on theory, experimentation, analysis and mechanisms of phototransferred thermoluminescence.

1.1 Phototransferred thermoluminescence

'Perfect is the enemy of good', so goes the aphorism, and in solids perfection does not exist. Intrinsic or extrinsic point defects in an insulator or semiconductor mar its structural and electronic regularity and create sites within the energy band gap where wave-like solutions of the Schrödinger equation do not exist. Some of these sites may have sufficient potential to trap and retain itinerant electrons. An electron restricted by binding potential to these types of imperfections will repeatedly attempt to detach from its trap. The release of trapped electrons can be facilitated by phonon or photon-aided transitions. Such thermally or optically related processes can be exploited to study the dynamics involving luminescent point defects. This is because the temperature-resolved glow peaks in an isometric temperature-intensity-wavelength spectrum denote the emptying of electron traps whereas the wavelength-dependent emissions relate to luminescence sites.

If the luminescence thermally stimulated from one electron trap is due to electrons optically re-distributed to it from another, that thermoluminescence is, in name, 'phototransferred'. In this sense, phototransferred thermoluminescence is a phonon-mediated temperature-and-wavelength resolved emission that arises at one electron trap due to a combination of phonon-aided and photon-mediated release of electrons from another electron trap. The phonon contribution at the source or donor electron trap is negligible if illumination is done at ambient temperature. However, since conventional thermoluminescence occurs owing to heating only, this process exclusively involves phonons. Photon-dependent electron release, that is, optical stimulation, is an inefficient process where signals need to be discriminated

doi:10.1088/978-0-7503-3831-8ch1

against background noise and electronically amplified at the photon counting stage. The contribution to the optically stimulated luminescence owing to the increase in illumination temperature is normally comparatively less. The additive effect of phonons to photon-aided release of electrons, otherwise termed thermal assistance, is therefore often obscured. Although phototransferred thermoluminescence and thermally-assisted optically stimulated luminescence are usually studied individually, the separation is artificial since the process of optical stimulation essential for phototransfer to occur is facilitated by thermal assistance. This book will therefore also cover optical and thermal processes that affect phototransfer in materials.

1.2 Thermoluminescence

In practical terms, phototransferred thermoluminescence is thermoluminescence owing to optical transfer of electrons between electron-trapping point defects. Thermoluminescence itself is a method for sensing changes in the concentration of charge-trapping defects in materials. When an irradiated material, usually an insulator or semiconductor, is heated at a well-defined rate, it glows and wanes in intensity successively. The shorthand for this phenomenon is thermoluminescence. Thermoluminescence can be recorded in terms of temperature only or as a wavelength resolved set of signals during controlled heating of the previously irradiated material. The ionizing radiation creates free electrons and holes whose radiative recombination at certain trapping sites is the signature that is thermoluminescence. If the measurements are restricted to recording the temperature-dependent set of signals, each resulting glow peak indicates the emptying of a particular electron trap. When the measurements are obtained as a spectrum, the wavelength encodes information about the recombination processes. By convention, the composition of the Fermi gas is confined to electrons but a discussion where the itinerant charge carriers were holes would be equally valid.

Figure 1.1 shows isometric plots of thermoluminescence measured from 25 °C to 400 °C from x-ray irradiated Al_2O_3:C,Mg, a polymorphic phase insulator. A prominent emission band centred near 410 nm at 150 °C stands out. Its extreme intensity suppresses any other fine features of the emission spectrum. When the heating is restricted to the first 100 °C, as illustrated in figure 1.1(b), other otherwise blurred features below 50 °C, which also correspond to the same emission band, become evident.

Figure 1.2 shows glow curves measured from two samples of α-Al_2O_3:C, here labelled A and B. Two types of responses are apparent. The glow curve of the type A sample consists of a dominant peak at 180 °C and secondary weaker-intensity components at either end. In comparison, the type B sample shows a similar result except for the lower intensity peak. The two samples are from the same batch and supplier, their differences notwithstanding.

The examples of figures 1.1 and 1.2 highlight several important aspects of thermoluminescence. One is that materials have a motley of point defects that act as electron traps and luminescence sites. Some are dominant and others less responsive to thermal stimulation. A glow curve necessarily reflects the dominant

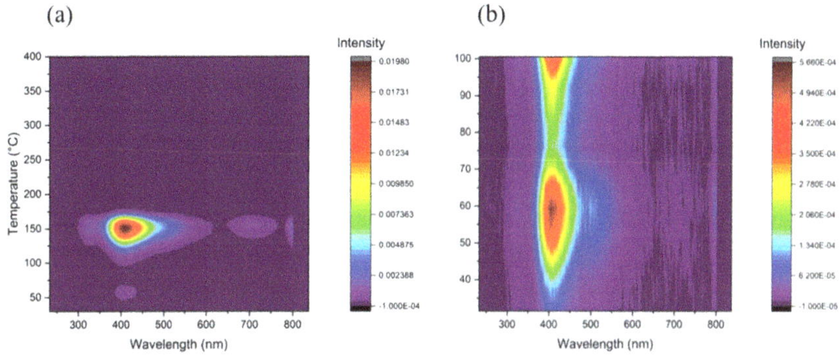

Figure 1.1. Contour maps of thermoluminescence measured at 10 °C min^{-1} from x-ray irradiated Al_2O_3:C, Mg. The signal recorded up to 400 °C is dominated by the emission near 150 °C which suppresses lower intensity contributions (a). When measurements are restricted to the first 100 °C and recorded after a delay of 400 s after irradiation, previously concealed emissions near 50 °C become clearer (b).

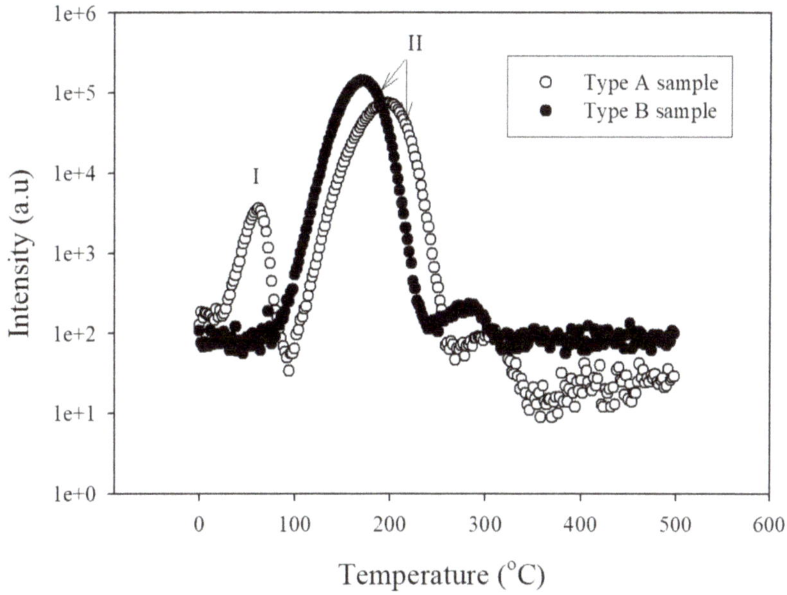

Figure 1.2. An overlay of thermoluminescence glow curves measured from two samples of α-Al_2O_3:C, one of which has shallow electron traps (open circles). The measurements were both made at 1 °Cs^{-1} after irradiation to 1 Gy. Reprinted from [1], Copyright (2015), with permission from Elsevier.

rather than all electron-trapping defects in a material. The sensitivity of the detection system affects how well a glow curve can be measured. As such, the peaks that compose a glow curve are nominal rather than exhaustive in number. Applications of thermoluminescence exploit its capacity to sense the presence of point defects in materials and its ready response to change in dose of irradiation used. The so-called 'thermally disconnected' or 'deep traps' are those that activate at temperatures beyond the heating limit of most contemporary instrumentation, that is, 500–700 °C. Another defining factor relates to the nature of point defects. Imperfections usually aggregate spatially [2], may lie close energy-wise and their presence may be random rather than systematic as is implied in figure 1.2. Thus, examples of types of glow peaks span single well-separated ones, others that overlap but where the components are discernible, and even collocated cases where the components are embedded to such an extent that the product speciously appears to be single. Examples of the latter, whose study lags well behind the other cases, are common in superluminous materials such as $SrAl_2O_4:Eu^{2+}$, Dy^{3+} (e.g. [3]). Indeed, as discussed elsewhere [4], the notion of a single isolated point defect may only be academic. For this reason, any analytical theory applicable to a discrete peak must be a limiting case of the complex cases outlined.

1.3 Models of thermoluminescence

The conceptual basis of thermoluminescence is that it is a thermodynamic process on a material of given properties under a set of principles. The interpretation and analysis of thermoluminescence can then be set within these three elements; namely, properties, principles and processes.

1.3.1 The tutorial one-trap one-recombination-centre approximation

An idealised model of thermoluminescence is based on a system of one electron trap and one recombination centre. This is shown in figure 1.3. Such a set-up is only a tutorial device and is necessarily inconsistent with defect properties of insulators or semiconductors because it fails to satisfy the triad elements listed earlier. Nevertheless, the utility of this model is that it offers a useful starting point to

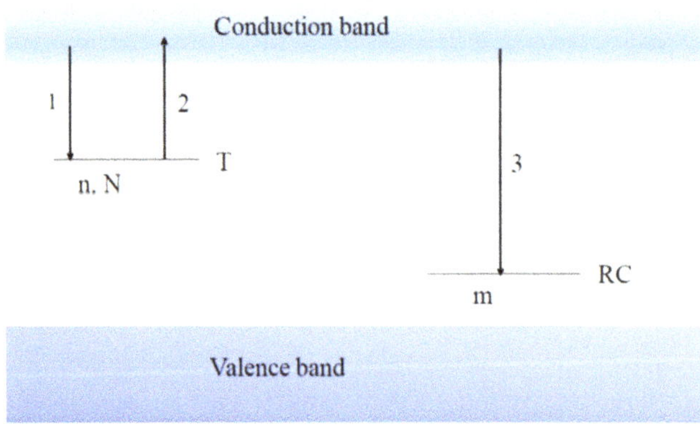

Figure 1.3. A simple one trap–one recombination centre model used to discuss thermoluminescence. Electrons moved to the electron trap (transition 1) can be thermally stimulated back to the conduction band (transition 2) from where transfer (transition 3) to a recombination centre (RC) can occur. The parameters N, n and m denote the concentration of electron traps, trapped electrons and holes in that order.

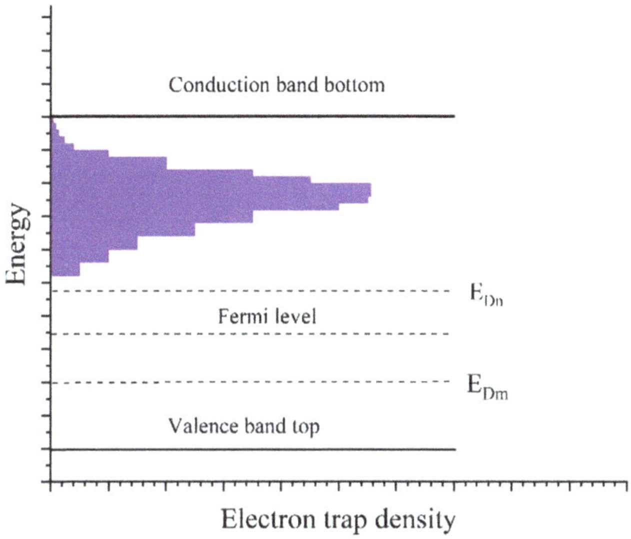

Figure 1.4. A schematic diagram of the energy against the electron trap density showing the likely distribution of electron trap levels for a hypothetical density of states.

explain facets of thermoluminescence. The fundamentals of thermoluminescence theory are discussed more expansively elsewhere [5–12] and our outline is abridged.

We assume a uniform distribution of electron-trapping levels for which the density of states is $D(E)$. This is shown as an energy-resolved spectrum in figure 1.4.

Since the system under consideration is above absolute zero of temperature, energy states beyond the Fermi energy are occupied. For convenience, one defines an energy cut-off point E_{Dn} where the probability $p(E)$ of thermal excitation from an electron trap to the conduction band exceeds the probability of recombination at a localised energy level that lies closer to the valence band [5]. A cut-off point E_{Dm} can be defined for holes in similar terms. With this understanding, the thermally-induced transfer of electrons from an electron trap to the conduction band and possible backscatter to the electron trap can be expressed in a more general way as

$$\frac{dn_c}{dt} = \int_{E_{Dn}}^{E_c} p(E)D(E)f(E)\mathrm{d}E - n_c A_n \int_{E_{Dn}}^{E_c} D(E)(1 - f(E))\mathrm{d}E - n_c A_{mn} \int_{E_{Dp}}^{E_F} D(E)(1 - f(E))\mathrm{d}E \tag{1.1}$$

$$\frac{dn}{dt} = n_c A_n \int_{E_{Dn}}^{E_c} D(E)(1 - f(E))\mathrm{d}E - \int_{E_{Dn}}^{E_c} p(E)D(E)f(E)\mathrm{d}E \tag{1.2}$$

where n_c (in m^{-3}) is the concentration of electrons in the conduction band, A_n (in $\mathrm{m}^3\,\mathrm{s}^{-1}$) is the probability of retrapping, A_{mn} (in $\mathrm{m}^3\,\mathrm{s}^{-1}$) is the probability of recombination and the function $f(E)$ gives the probability that an energy state is occupied. Equations (1.1) and (1.2) cannot be simplified further unless the function $D(E)$ is known. To circumvent this requirement, the likely distribution of localised levels can be replaced by the simplifying assumption that there are two discrete energy level types which then makes $D(E)$ finite. One energy level represents an electron trap and the second, a recombination centre. In this way, the meaning of the integrals in equations (1.1) and (1.2) become immediately clear as follows:

$$\int_{E_{Dn}}^{E_c} D(E)f(E)\mathrm{d}E \equiv n \tag{1.3}$$

$$\int_{E_{Dn}}^{E_c} D(E)(1 - f(E))\mathrm{d}E \equiv N - n \tag{1.4}$$

$$\int_{E_{Dp}}^{E_F} D(E)(1 - f(E))\mathrm{d}E \equiv m \tag{1.5}$$

where n is the concentration of electrons at the electron trap, $(N - n)$ is the concentration of empty electron traps and m the concentration of holes at the recombination centre. Substituting equations (1.3)–(1.5) in equations (1.1) and (1.2), we find that

$$\frac{dn_c}{dt} = np - n_cA_n(N - n) - n_cmA_{mn} \tag{1.6}$$

$$\frac{dn}{dt} = n_cA_n(N - n) - np \tag{1.7}$$

and by definition, aided by dimensional analysis, the last term of equation (1.1) is

$$-\frac{dm}{dt} = n_cmA_{mn} \tag{1.8}$$

Equations (1.6)–(1.8) are derived by restricting the thermally stimulated luminescence to the release of trapped electrons as should be implicit from figure 1.4. A similar set of expressions can be obtained by assuming that the signal is obtained due to the release of holes. There are a number of important points that emerge from this formulation, the analysis of which are essential in addressing the features of phototransferred thermoluminescence:

a) Equations (1.6)–(1.8) are obtained by invoking many simplifying assumptions into the underlying theory.
b) Thermoluminescence measurement apparatus record or convert the emission to a temperature- not time-resolved set of emissions. Therefore each one of the variables n, n_c and m are functions of functions of temperature, that is, they are functionals.
c) The system of equations (1.1)–(1.2) or (1.6)–(1.8) refers to a discrete set of localised energy levels. In that sense, thermoluminescence theory, including the foundational case of one-trap one-recombination-centre, is only an approximation theory.

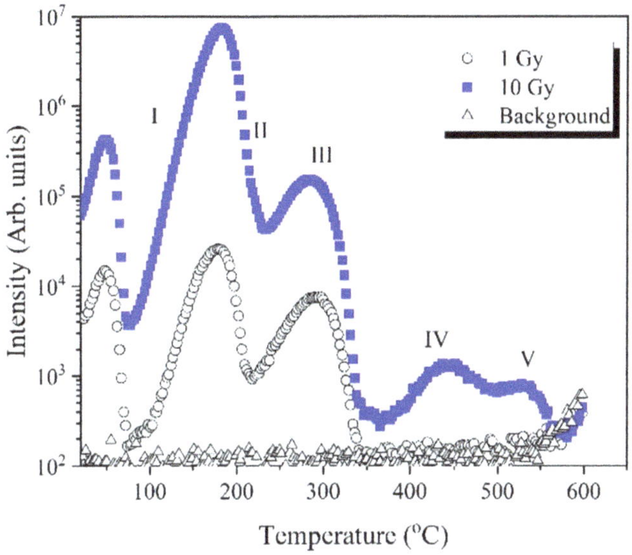

Figure 1.5. Glow curves of BeO corresponding to 1 and 10 Gy beta dose. The background signal is shown for completeness. The plot shows that peaks beyond 400 °C are only induced following the larger dose. Reprinted from [13], with the permission of AIP Publishing.

d) Reducing a distribution of energy levels to a set of discrete electron traps (or recombination centres) dispenses with reality. The consequence of this is that analytical methods based on equations (1.6)–(1.8) properly describe only dominant emissions.

As a case in point for comment (d) above, consider figure (1.5) showing glow curves measured from BeO following irradiation to 1 and 10 Gy [13]. The glow curve obtained after irradiation to 1 Gy has three nominal peaks. However, when the dose is changed to 10 Gy, two additional peaks beyond 400°C come up. These latter two are too weak to be distinct against the backdrop of the much intense first three peaks when a low dose of 1 Gy is used. Even so, it is only possible to pick out the differences when the dynamic range is compressed on a semi-logarithmic plot as done.

1.3.2 Beyond the instructive: multiple localised states

Experiments show that materials contain an extensive array of point defects that can serve as electron traps and luminescence sites [2, 14]. This is clearly apparent in thermoluminescence spectra, say figure 1.6, where electron traps correspond to glow curves and luminescence sites to emission bands. The energy band model of figure 1.3 can be modified to achieve a more realistic description of the system by including so-called 'deep' or 'thermally disconnected' electron traps. These are not emptied during routine heating of the material but rather act as electron reservoirs. Such point defects can also serve as competitors for charge that would otherwise be transferred to other electron traps. Deep electron traps are sometimes relevant for

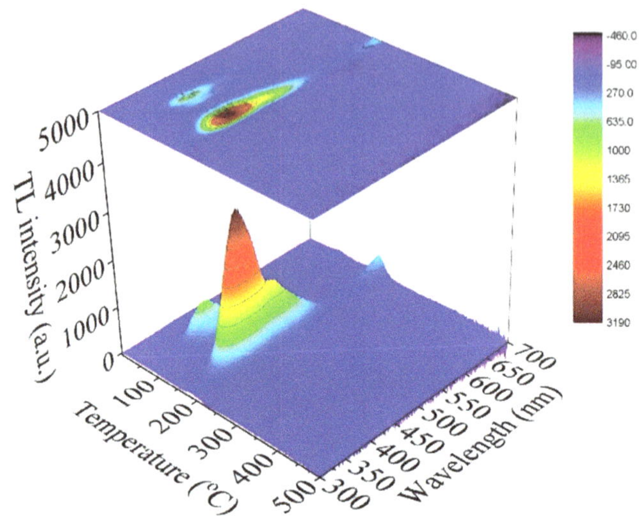

Figure 1.6. A thermoluminescence emission spectrum measured from Al_2O_3:C,Mg after beta irradiation to 10 Gy. The sample was annealed at 900 °C before use. Reprinted from [15], Copyright (2018), with permission from Elsevier.

discussion of phototransferred thermoluminescence. If the concentration of electrons retained at such deep electron traps is n_d, the system of equations in question becomes

$$\frac{dn_c}{dt} = np - n_c A_n(N - n) - n_c m A_{mn} \tag{1.9}$$

$$\frac{dn}{dt} = n_c A_n(N - n) - np \tag{1.10}$$

$$\frac{dn_d}{dt} = A_d(N_d - n_d) \tag{1.11}$$

$$-\frac{dm}{dt} = n_c m A_{mn} \tag{1.12}$$

where A_d is the electron-trapping probability at the deep electron trap whose total concentration is N_d. Other extensions to the simple model include treatments intended to address the possibility of localised transitions including quantum tunnelling [16] or charge hopping [17].

1.4 Calculational methods

Equations (1.6)–(1.8) are a set of coupled non-linear differential equations and as such do not have an exact solution. In order to obtain a single expression that can be used to describe a glow peak, further simplifications are necessary. Assuming that electrons do not accumulate in the conduction band such that their concentration here is stable, we can write that

$$n_c \ll n \tag{1.13}$$

hence,

$$\frac{dn_c}{dt} \ll \frac{dn}{dt} \tag{1.14}$$

Another requirement to fold into the discussion is that the Gibbs free energy for the system cannot be minimised unless the constraint $\sum Q_j n_j = 0$ where Q_j denotes charge and n_j their number, is satisfied. This means that the concentrations of electrons and holes within the volume of the sample under study must balance, that is

$$n_c + n = m \tag{1.15}$$

Combining equations (1.13) and (1.15) with equations (1.6)–(1.8), one obtains an expression for n_c as

$$n_c = \frac{np}{[(N - n)A_n + mA_{mn}]} \tag{1.16}$$

Therefore, the terms $-dm/dt$ or $-dn/dt$, either of which expresses the time-dependence of the luminescence intensity, now read

$$I(t) = -\frac{dm}{dt} = n_n m A_{mn} \tag{1.17}$$

This can then be expressed as

$$I(t) = -\frac{dm}{dt} = \frac{pn^2}{A_m m + A_n(N - n)} \tag{1.18}$$

Equation (1.18) cannot be used as is to analyse thermoluminescence glow peaks because, as stated earlier, some of its variables are functionals. Since thermoluminescence is recorded as a set of temperature-resolved peaks, the intensity ought to be a function of temperature only. To achieve this, we need to know how to describe the thermoluminescence intensity for the particular process involved. These processes are identified by the so-called order of kinetics. These descriptions are assigned to processes depending on whether electrons thermally released from electron traps mostly scatter back to the origin trap or not. Kinetic analysis, which encompasses techniques for evaluating other parameters other than the order of kinetics, is a touchstone of thermoluminescence theory. The basics for first and second order kinetics go back to Randall and Wilkins [18] and Garlic and Gibson [19] and have been reviewed extensively elsewhere [5–12].

1.4.1 First order kinetics

First order processes, that is where $-dn/dt \; \alpha \; n$ arise when instances of back-scattering are negligible, that is

$$(N - n)A_n \ll mA_{mn} \tag{1.19}$$

which, using equation (1.18) and invoking the quasi-equilibrium approximation, i.e. $n_c \approx 0$, means that

$$-\frac{dn}{dt} = pn \tag{1.20}$$

Since p (in s^{-1}), the probability of thermal excitation (originally written as $p(E)$ in equation (1.1)), is given by $p = s \exp(-E/kT)$ [5, 6], we can re-write equation (1.20) as

$$-\frac{dn}{dt} = ns \exp(-E/kT) \tag{1.21}$$

where the factor s is understood as the frequency with which a trapped electron attempts to detach from its binding potential. By assuming a linear heating rate β, the temperature dependence of the thermoluminescence intensity $I(T)$ for first order processes is obtained as

$$I(T) = n_o s \exp(-E/kT) \exp\left[-\frac{s}{\beta}\int_{T_o}^{T} \exp(-E/k\theta)\mathrm{d}\theta\right] \tag{1.22}$$

where n_o is the initial concentration of trapped electrons.

The primary experimental features of a first order peak are as follows:

(a) Qualitatively, the peak appears to be asymmetric being wider at its rising edge than at its higher temperature end. This characteristic is better evident for isolates.

(b) The position of a first order peak is insensitive to the concentration of trapped electrons. Its position remains stable, allowing for statistical scatter, despite change in the concentration of trapped electrons n caused by change in excitation dose, slight partial heating up to some point on its rising edge, or owing to fading of the peak.

1.4.2 Second order kinetics

Second order kinetics where $-dn/dt \; \alpha \; n^2$ describes processes where retrapping is recurrent. Therefore,

$$(N - n)A_n \gg nA_{mn} \tag{1.23}$$

Assuming $n \ll N$, equation (1.18) simplifies to

$$I(t) = -\frac{dn}{dt} = \frac{n^2 p}{(A_n/A_{mn})(N - n)} \approx \frac{n^2 p}{(A_n/A_{mn})N} \tag{1.24}$$

Proceeding as before, assuming linear ramping at a rate β, the change of the thermoluminescence intensity with temperature for second order kinetics is found as

$$I(T) = n_o \frac{s}{N} \exp(-E/kT)\left[1 + \frac{n_o s}{\beta N}\int_{T_o}^{T} \exp(-E/k\theta)\mathrm{d}\theta\right]^{-2} \quad (1.25)$$

What marks a second order peak besides its symmetry is that its position decreases with trap occupancy or irradiation dose. Conversely, the peak will appear to shift to higher temperatures with fading. Although such experimental properties have mathematical basis, it is inadvisable to pick out and draw a conclusion about order of kinetics from a single test alone. Instead, the order of kinetics must be corroborated by qualitative or quantitative kinetic analysis. Methods for this are described in detail in suitable texts [5–8, 12].

1.4.3 Other descriptions of thermoluminescence processes

1.4.3.1 General order kinetics

First and second order kinetics are the only canonical descriptions of thermoluminescence processes. There are other means to address experimental results that are inconsistent with either first or second order kinetics. In a form due to Rasheedy [20], general order processes are expressed as

$$I(t) = -\frac{dn}{dt} \equiv \left(\frac{n^b}{N^{b-1}}\right)s \exp(-E/kT) \quad (1.26)$$

If the exponent b is not equal to either 1 or 2 denoting first or second order kinetics, it will be intermediate between these two. The meaning of b outside these two extremes is moot. For a constant heating rate, the signal intensity is expressed as a function of temperature in this case as

$$I(T) = n_o^b s N^{1-b} \exp(-E/kT)\left[1 + \frac{s(b-1)(n_o/N)^{b-1}}{\beta}\int_{T_o}^{T} \exp(-E/k\theta)\mathrm{d}\theta\right]^{\frac{-b}{b-1}} \quad (1.27)$$

1.4.3.2 Mixed order model

The mixed order model arises when deep electron traps, that is, those that can only be activated at elevated temperatures, are included in the model of figure 1.3. In this case, the charge neutrality condition $m = n$ does not apply because of restrictions on energy minimisation required by the Gibbs free energy. Therefore, instead of using equation (1.15) one writes

$$m = n + n_c + |c| \quad (1.28)$$

where $|c|$ is a constant. If $n_c \approx 0$, by invoking the quasi-equilibrium approximation, $m = n + |c|$. If we then set $A_n = A_{mn}$ and substitute these expressions in equation (1.18), we find

$$I(T) = -\frac{dn}{dt} = \left(\frac{s}{|c| + N}\right)n \quad (n + |c|)\exp(-E/kT) \tag{1.29}$$

If instances of retrapping are frequent, that is, $A_{mn}m \ll A_n(N - n)$ and $n \ll N$, we find

$$I(T) = -\frac{dn}{dt} = \frac{sA_{mn}}{A_nN}n \quad (n + |c|)\exp(-E/kT) \tag{1.30}$$

The solution of equation (1.30) for a constant heating rate is

$$I(T) = \frac{S'c^2\Gamma_n \exp\left[\left(S'|c|/\beta\right)\int_{T_0}^{T} \exp(-E/k\theta)d\theta\right]\exp(-E/kT)}{\left\{\exp\left[\left(S'|c|/\beta\right)\int_{T_0}^{T} \exp(-E/k\theta)d\theta\right] - \Gamma_n\right\}^2} \tag{1.31}$$

where $\Gamma_n = n_o/(n_o + |c|)$ and n_o have the same meaning as defined earlier.

Kirsh [10] notes that if either of equations (1.29) or (1.30) is written as

$$I(T) = S'n \quad (n + |c|)\exp(-E/kT) \tag{1.32}$$

one obtains

$$I(T) = S'n \quad (n + |c|)\exp(-E/kT) + S'n^2\exp(-E/kT) \tag{1.33}$$

where $S' = s/(N + |c|)$. Equation (1.33) reduces to first order when $c \gg n$ or second order when $c \ll n$. This explains the choice of the term 'mixed order' to refer to this model.

1.5 Defects and disorder in solids

The discussion thus far has been founded on the principle that thermoluminescence ensues from and is affected by the presence of imperfections which introduce energy levels within the energy band gap. The existence of these localised levels is rationalised to be a consequence of modifying the density of states. The unstated assumption in that description is that the material is crystalline. This is not true for all solids. We can gain more information on the origin of localisation and other effects on the energy band gap by considering not just point defects but disorder in materials.

The allowed energies of a conduction electron in a perfect lattice can be found by solving the time-independent Schrödinger equation,

$$\mathscr{H}\,|\Psi_n\rangle = E_n\,|\Psi_n\rangle \tag{1.34}$$

where $\mathscr{H}$ is the Hamiltonian of the system and E_n the energy eigenvalues of state $|\Psi_n\rangle$. The potential $V(r)$ in the Hamiltonian is taken as that which the electron sees at each lattice point. For convenience each such lattice point is treated as a point ion so that $V(r)$ becomes a periodically varying potential. When defects are introduced in the lattice, the periodicity is disturbed since the potential at the defect sites differs from that at the host crystal ions. The defect potential $U(r)$, say, perturbs the system which necessitates modifying the Hamiltonian hence equation (1.34) is re-written as

$$[\mathscr{H}_o + U(r)] \mid \Psi_n\rangle = E_n \mid \Psi_n\rangle \tag{1.35}$$

where $\mathscr{H}_0$ represents the unperturbed Hamiltonian. The solutions to equation (1.35) show that states are split off from the band under consideration (e.g. see [21]). If the extra potential $U(r)$ is positive as when electrons are repelled by the defects, the states are split off the uppermost band. If the defect tends to attract electrons then the negative $U(r)$ leads to break-off on the lowermost bands. The wave function associated with such states are localised in space at the defect [21]. The presence of any disorder therefore introduces localisation in both defect-containing crystalline materials as well as in disordered solids. How these levels affect the luminescence is not well understood for all defects. However, it is appreciated that changes in the occupancy of the various localised energy states, owing to electronic transitions from state to state, are essential for all luminescence phenomena to occur. Although thermoluminescence has been discussed mostly in these terms, this is not the only model that can account for this phenomenon.

Another consequence of the introduction of disorder is broadening of the energy bands as illustrated in figure 1.7. This causes tailing of states into the band gap as demonstrated by Weaire and Thorpe [22]. The presence of band tail states is sometimes also relevant for discussion of luminescence in both crystalline and disordered materials and has repeatedly been cited to explain thermoluminescence and optically stimulated luminescence in feldspar (e.g. [24, 25]).

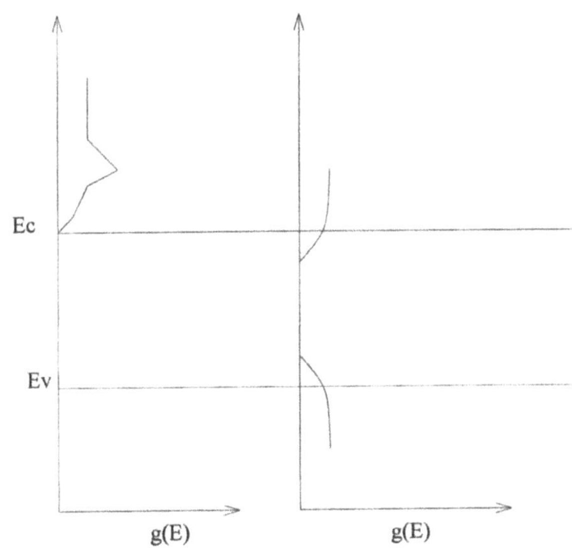

Figure 1.7. The density of states $g(E)$ for a crystalline material and one containing a small amount of disorder. Whereas the crystalline material exhibits van Hove singularities, when some disorder is introduced, the singularities become smeared out. Instead, some localised states tail into the band gap. After Sidebottom [23].

1.5.1 Defect pair model

The defect pair model is well established in the study of luminescence in alkali halides (e.g. [26]). The context is the generation of point defects by radiolysis [2]. In this account, luminescence arises due to the recombination of defect pairs consisting of a halogen ion vacancy (an F^+ centre) and an H centre (a neutral crowdion interstitial). This model was used by Itoh *et al* [27] to discuss thermoluminescence and optically stimulated luminescence in quartz. A key feature of the application to quartz is the existence of the species [XO_4] that stabilise mobile alkali ions. Vaccaro *et al* [28] investigated the identity of X and forwarded Ge as a possibility. A recurrent theme in the study of thermoluminescence of quartz is the interpretation of its spectra (e.g. [29]). In this regard, Williams and Spooner [30] used the defect pair model to explain the spectral properties of some thermoluminescence peaks in quartz. In applications to optically stimulated luminescence, Williams *et al* [31] used the defect pair model to account for the apparent appearance of cascading steps in time in decay curves measured over an extended duration. Williams and Spooner [32] discussed a configurational coordinate model for optically stimulated luminescence in quartz using the defect pair model and applied it to explain thermal assistance in this material. The defect pair model has also been used to explain thermoluminescence and phototransferred thermoluminescence in tanzanite [33].

1.6 Non-radiative transitions

Thermal or optical stimulation of electrons cannot always result in radiative transitions. To expect so would be to wrongly impart a classical characteristic to what are, in reality, probabilistic events. Not all recombinations lead to luminescence. Non-radiative transitions may be explained using several models and here we single out intra-band transitions and the configurational coordinate model.

1.6.1 Intra-band transitions

Non-radiative transitions are an important element of temperature-dependent effects in luminescence. These processes cause a decrease in the luminescence intensity with measurement temperature. This 'thermal quenching' has been extensively discussed for various luminescence phenomena [5, 6, 34–36]. An example of intra-band transitions leading to non-radiative outcomes is the so-called Schön–Klasens model sketched in figure 1.8. The basic notion is that holes are released from radiative recombination centres at elevated temperatures and transferred via the valence band to non-radiative recombination centres.

Electrons recombining at non-radiative recombination centres are thought to dissipate their energy as phonons. Thermal quenching can be interpreted as a direct result of a decrease in the number of radiative recombination centres or an increase in the concentration of non-radiative ones. When luminescence occurs, it does so

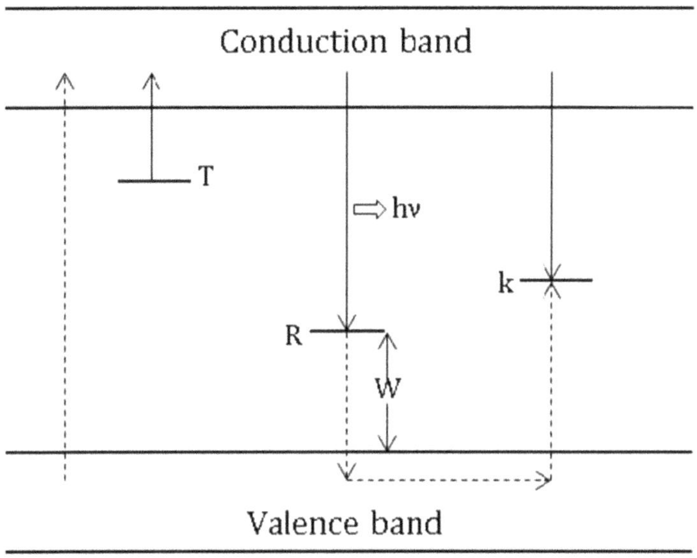

Figure 1.8. The energy band scheme for the Schön–Klasens model used to describe thermal quenching. The dashed line between delocalised energy bands denotes ionization. The electron trap is labelled T whereas the radiative recombination centre at an energy W above the valence band is labelled R. Holes are transferred from the recombination centre R to a non-radiative centre (k) via the valence band.

from radiative recombination of electrons and holes at the radiative recombination centres. The efficiency η of the recombination process is given by

$$\eta = \frac{1}{1 + C\exp(-W/kT)} \tag{1.36}$$

where T is the absolute temperature and the constant $C = \nu\tau_{\text{rad}}$ where τ_{rad} is the radiative lifetime at 0 K and ν is the frequency factor for non-radiative transitions [37].

The value of W can be abstracted from a plot of luminescence lifetimes against annealing temperature. An example of such a plot is shown in figure 1.9. These types of measurements are done using time-resolved optical stimulation [35]. As an illustration, experiments made on quartz have been discussed with reference to the scheme of figure 1.10.

In this scheme, annealing causes the transfer of holes from a non-radiative recombination centre R to and between radiative ones L_H, L_L and L_S, which are each associated with lifetimes τ_H, τ_L and τ_S where $\tau_H > \tau_L > \tau_S$ [35]. Luminescence lifetimes change from a higher value τ_H to a lower one τ_L with annealing temperature in a process consistent with the transfer of holes between recombination centres L_H, L_L and L_S. The use of three types of recombination centres in this example is empirical and refers to the quartz studied which showed three distinct

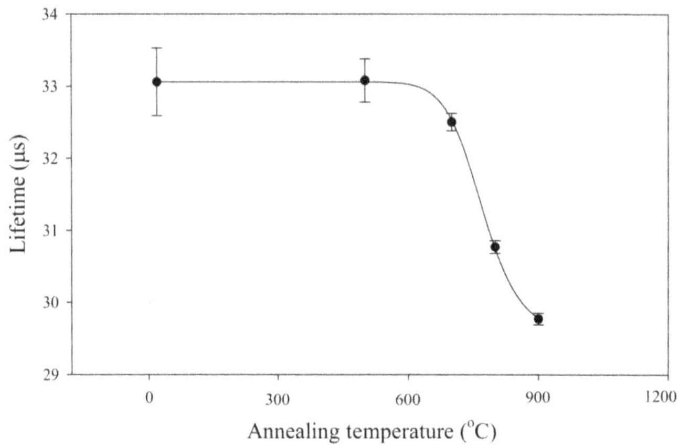

Figure 1.9. Dependence of luminescence lifetimes on annealing temperature for measurements made at 25 °C on quartz from Mato Grosso, Brazil. The luminescence was pulse stimulated using 470 nm LEDs. Reprinted from [38], Copyright (2008), with permission from Elsevier.

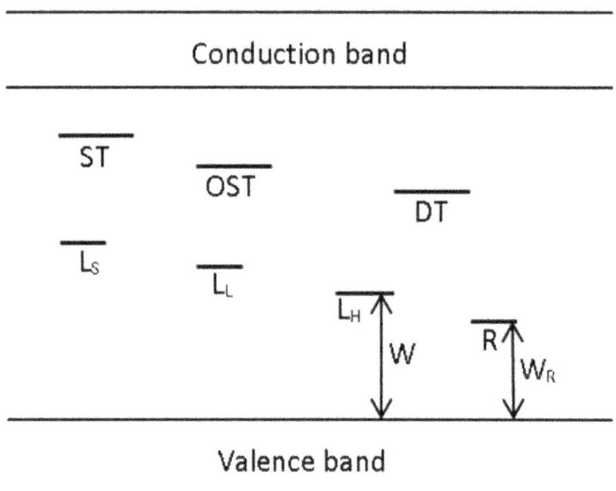

Figure 1.10. The energy band scheme used to explain the effect of annealing on luminescence lifetimes in quartz used a way to evaluate W. The recombination centres are denoted L_H, L_L and L_S and the non-radiative centre as R. The symbols ST, OST and DT refer to the shallow electron traps, optically sensitive traps and deep traps. Reprinted from [39], Copyright (2002), with permission from Elsevier.

lifetimes. The change of mean lifetimes τ with annealing temperature T is empirically expressed as

$$\tau = \tau_L + \frac{(\tau_H - \tau_L)}{1 + C\exp(-W/kT)} \tag{1.37}$$

where W is as explained earlier. The line through the data in figure 1.9 is a fit of equation (1.37), which offers a means to find W.

1.6.2 Configurational coordinate model

Non-radiative transitions can also be accounted for using configurational coordinate diagrams. Such a diagram shows the energy of a system with respect to the displacement or offset from the equilibrium position of the nuclei (atoms or ion) of which the system is composed. The utility of configurational coordinate diagrams in explaining photo absorption, and photoemission (or luminescence) in systems subject to strong electron-lattice coupling that appears in localised systems has been discussed elsewhere (e.g. [40]).

Figure 1.11 shows configurational coordinate adiabatic potential energy curves corresponding to strong electron-lattice coupling. The curves intersect at point P. Figures 1.11(a) and (b) correspond to maximum absorption. The probability that an electron in an excited state will relax is described by a multiphonon relaxation term (transition 1; figure 1.11(a)) from state B to state C from where emission (transition 2) can occur. Another multiphonon relaxation to state A (transition 3) follows and this corresponds to the equilibrium position $q_o^{(g)}$ of the ground state. If the energy of state B exceeds that of state P, the electron can move to the vibrational state corresponding to point P (transition 1; figure 1.11(b)). In this case, point P is degenerate to both the ground and the excited states. Because of this, the system forgoes any occupation of state C leaving this state empty. Therefore, the probability that relaxation will occur via phonon states of initial state *g* (transition 5; figure 1.11(b)) exceeds any likelihood of this process occurring via state *e* and hence no luminescence is emitted. An electron at the minimum of state *e* can absorb an energy ΔE to transit to point P from where non-radiative transition occurs. These processes are summarised below

$$\frac{1}{\tau} = \frac{1}{\tau_{\text{rad}}} + \gamma \coth\left(\frac{\hbar\omega}{kT}\right) + \nu \exp\left(\frac{-\Delta E}{kT}\right) \tag{1.38}$$

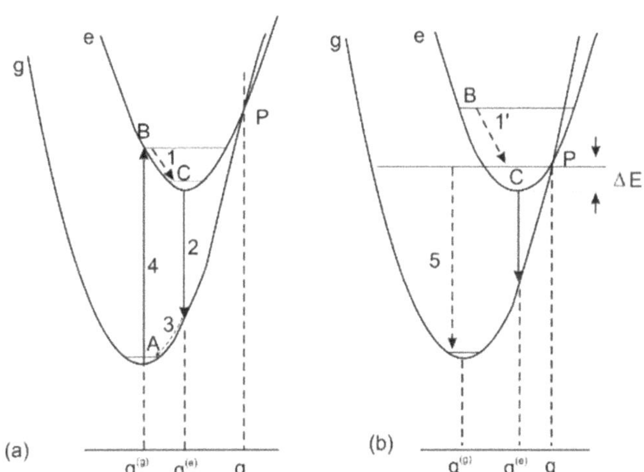

Figure 1.11. Configurational coordinate diagrams used to account for radiative transitions (a) and thermal quenching (b). Reprinted from [41], Copyright (2015), with permission from Elsevier.

where τ is the lifetime of the excited state at the recombination centre, τ_{rad} is the lifetime at 0 K, γ is a constant, ω is the phonon vibration frequency, $\hbar$ is the reduced Planck's constant, k is Boltzmann's constant, ΔE and ν are, respectively, the activation energy and frequency factor for non-radiative processes [2].

Methods for evaluating the activation energy of thermal quenching include those based on the influence of heating rate on thermoluminescence intensity (e.g. [8]) and others that exploit the temperature dependence of luminescence lifetimes as measured in time-resolved luminescence [35].

1.6.2.1 Mott–Seitz model

The Mott and Seitz model of thermal quenching uses a flat energy band diagram by referencing localised energy levels to the elements of a configurational coordinate diagram. We explain the model by showing how it was adapted by Pagonis *et al* [41] to account for thermal quenching in quartz. A schematic of the thermal quenching model used by Pagonis *et al* [41] is displayed in figure 1.12. The stimulation here is optical. Transitions within the recombination centre relate to the configurational coordinate description. It is assumed that some electrons stimulated from an electron trap to the conduction band (transition 1) move to an excited state of the recombination centre (transition 2) with a probability A_{CB} (in m^3 s^{-1}). As in the configurational coordinate model, electrons in the excited state (shown as level 2) can make a direct radiative transition to the ground state or do so indirectly (arrow 6) from another excited state (level 3), which is accessible to both levels 2 and the ground state. The direct radiative transition proceeds with probability A_R (in s^{-1}). Pagonis *et al* [41] reasoned that the transfer of electrons from excited state level 2 to level 3 occurs with probability $A_{NR}\exp(-\Delta E/kT)$ where A_{NR} is the probability of non-radiative transitions.

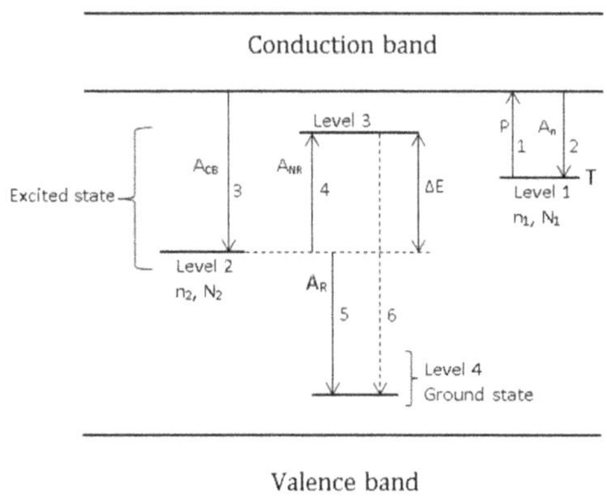

Figure 1.12. The energy band diagram used to describe thermal quenching in the kinetic model adaptation of the Mott–Seitz model. Concerning electron movement, some of the electrons optically stimulated to the conduction band are re-trapped at the electron trap (transition 2) with probability A_n in (m^3 s^{-1}). Reprinted from [41], Copyright (2010), with permission from Elsevier.

The set of equations describing the various processes are then written as

$$\frac{dn_1}{dt} = -n_1P + A_n n_c(N_1 - n_1) \quad (1.39)$$

$$\frac{dn_2}{dt} = -A_R n_2 - n_2 A_{NR} \exp\left(-\Delta E/kT\right) + A_{CB} n_c(N_2 - n_2) \quad (1.40)$$

$$\frac{dn_c}{dt} = -n_c(N_1 - n_1)A_n + n_1P - A_{CB} n_c(N_2 - n_2) \quad (1.41)$$

where N_1 (m^{-3}) is the concentration of electron traps and n_1 (m^{-3}) the concentration of trapped electrons. For convenience, levels 2 and 3 are each treated as an electron trap. Thus, N_2 (m^{-3}) is the concentration of electron-trapping excited states and n_2 (m^{-3}) is the corresponding concentration at these electron traps. The instantaneous concentration of electrons in the conduction band is represented as n_c (in m^{-3}). The luminescence intensity is therefore expressed as

$$I(t) = A_R n_2 \quad (1.42)$$

Implicit in equation (1.42) is the fact that the luminescence intensity changes in time in correspondence to that of n_2. The effect of thermal quenching on the luminescence process is encapsulated in equation (1.40).

1.7 Overview

When an irradiated material is partially heated to deplete an electron trap as a first step in the measurement procedure for phototransferred thermoluminescence, the expectation is that it will be refilled by phototransfer. Thus, phototransfer, where the agent is light, simply re-distributes previously trapped electrons in electrons traps. The regeneration of a glow peak in this way is, however, only a fortunate exception. It is more common for phototransferred thermoluminescence to appear when several peaks, not just a single peak, have been removed. Indeed, only some peaks in a glow curve reappear under phototransfer. Competition effects—meaning processes which either enhance or depress the intensity of phototransferred thermoluminescence—sometimes cause signals from donor peaks to counterintuitively increase in intensity. The study of phototransferred thermoluminescence poses many questions that defy its conceptual formulation. This book will attempt to answer some, pose some more, and in so doing hopefully generate interest in this advancing topic.

References

[1] Chithambo M L, Nyirenda A N, Finch A A and Rawat N S 2015 Time-resolved luminescence and spectral emission features of α-Al_2O_3:C *Physica B: Condens. Matter.* **473** 62–71

[2] Agullo-Lopez F, Catlow C R A and Townsend P D 1988 *Point Defects in Materials* (London: Academic)

[3] Chithambo M L, Wako A H and Finch A A 2017 Thermoluminescence of $SrAl_2O_4:Eu^{2+}$, Dy^{3+}: kinetic analysis of a composite-peak *Radiat. Meas.* **97** 1–13

[4] Townsend P D, Wang Y and McKeever S W S 2021 Spectral evidence for defect clustering: relevance to radiation dosimetry materials *Radiat. Meas.* **147** 106634

[5] Chen R and McKeever S W S 1997 *Thermoluminescence and Related Phenomena* (Singapore: World Scientific)

[6] McKeever S W S 1985 *Thermoluminescence of Solids* (Cambridge: Cambridge University Press)

[7] Furetta C *Handbook of Thermoluminescence* (Singapore: World Scientific)

[8] Pagonis V, Kitis G and Furetta C 2006 *Numerical and Practical Exercises in Thermoluminescence* (Berlin: Springer)

[9] Bos A J J 2007 Theory of thermoluminescence *Radiat. Meas.* **41** S45–56

[10] Kirsh Y 1992 Kinetic analysis of thermoluminescence *Phys. Stat. Sol.* **129** 15–48

[11] McKeever S W S and Chen R 1997 Luminescence models *Radiat. Meas.* **27** 625–61

[12] McKeever S W S 2022 *A Course in Luminescence Measurements and Analyses for Radiation Dosimetry* (Hoboken, NJ: Wiley)

[13] Chithambo M L and Kalita J M 2021 Phototransferred thermoluminescence of BeO: time-response profiles and mechanisms *J. Appl. Phys.* **130** 195101

[14] Tilley R J D 2008 *Defects in Solids* (Hoboken, NJ: Wiley)

[15] Kalita J and Chithambo M L 2018 The effect of annealing and beta irradiation on thermoluminescence spectra of α-Al_2O_3:C,Mg *J. Lumin.* **196** 195–200

[16] Pagonis V 2019 Recent advances in the theory of quantum tunneling for luminescence phenomena *Advances in Physics and Applications of Optically and Thermally Stimulated Luminescence* ed R Chen and V Pagonis (Singapore: World Scientific)

[17] Kalita J M and Chithambo M L 2017 A comparative study of the dosimetric features of α-Al_2O_3:C,Mg and α-Al_2O_3:C *Radiat. Prot. Dosim.* **177** 261–71

[18] Randall J T and Wilkins M H F 1945 Phosphorescence and electron traps *Proc. Roy. Soc.* A **184** 366

[19] Garlick G F J and Gibson A F 1948 The electron trap mechanism of luminescence in sulphide and silicate phosphors *Proc. R. Soc. Lond.* **A60** 574

[20] Rasheedy M S 1993 A new method for obtaining the order of kinetics and the activation energy of the thermoluminescence glow peak *J. Phys. Condens. Matter* **5** 633

[21] Elliot S R 1990 *Physics of Amorphous Materials* (Hong Kong: Longman)

[22] Weaire D and Thorpe M F 1971 Electronic properties of an amorphous solid. I. A simple tight-binding theory *Phys. Rev.* **B4** 2508

[23] Sidebottom D L 2012 *Condensed Matter and Crystalline Physics* (Cambridge: Cambridge University Press)

[24] Jain M and Ankjærgaard C 2011 Towards a non-fading signal in feldspar: insight into charge transport and tunnelling from time-resolved optically stimulated luminescence *Radiat. Meas.* **46** 292–309

[25] Riedesel S, King G E, Kumar Prasad A, Kumar R, Finch A A and Jain M 2019 Optical determination of the width of the band-tail states, and the excited and ground state energies of the principal dosimetric trap in feldspar *Radiat. Meas.* **125** 40–51

[26] Itoh N and Stoneham A M 2001 *Materials Modification by Electronic Excitation* (Cambridge: Cambridge University Press)

[27] Itoh N, Stoneham D and Stoneham A M 2002 Ionic and electronic processes in quartz: mechanisms of thermoluminescence and optically stimulated luminescence *J. Appl. Phys.* **92** 5036–44
[28] Vaccaro G, Panzeri L, Paleari S, Martini M and Fasoli M 2017 EPR investigation of the role of germanium centres in the production of the 110 °C thermoluminescence peak in quartz *Quat. Geochronol.* **39** 99–104
[29] Rendell H M, Townsend P D, Wood R A and Luff B J 1994 Thermal treatments and emission spectra of TL from quartz *Radiat. Meas.* **23** 441–9
[30] Williams O M and Spooner N A 2018 Defect pair mechanisms for quartz intermediate temperature thermoluminescence bands *Radiat. Meas.* **108** 41–4
[31] Williams O M, Spooner N A, Smith B W and Moffatt J E 2018 Extended duration optically stimulated luminescence in quartz *Radiat. Meas.* **119** 42–51
[32] Williams O M and Spooner N A 2020 Quartz optically stimulated luminescence configurational coordinate model *Radiat. Meas.* **132** 106259
[33] Chithambo M L and Folley D E 2020 Dosimetric features, kinetics and mechanisms of thermoluminescence of tanzanite *Physica B: Condens. Matter.* **598** 412435
[34] Bøtter-Jensen L, McKeever S W S and Wintle A G 2003 *Optically Stimulated Luminescence Dosimetry* (Amsterdam: Elsevier)
[35] Chithambo M L 2018 *An Introduction to Time-resolved Optically Stimulated Luminescence* (San Rafael, CA: Morgan & Claypool Publishers)
[36] Henderson B and Imbusch G F 1989 *Optical Spectroscopy of Inorganic Solids* (Oxford: Oxford University Press)
[37] Chithambo M L 2007 The analysis of time-resolved optically stimulated luminescence. II: Computer simulations and experimental results *J. Phys. D: Appl. Phys.* **40** 1880–9
[38] Chithambo M L, Ogundare F O and Feathers J 2008 Principal and secondary luminescence lifetime components in annealed natural quartz *Radiat. Meas.* **43** 1–4
[39] Galloway R B 2002 Luminescence lifetimes in quartz: dependence on annealing temperature prior to beta irradiation *Radiat. Meas.* **35** 67–77
[40] Shinozuko Y 2021 Configurational coordinate diagram *Electron–Lattice Interactions in Semiconductors* (Singapore: Jenny Stanford Publishing)
[41] Pagonis V, Ankjærgaard C, Murray A S, Jain M, Chen R, Lawless J and Greilich S 2010 Modelling the thermal quenching mechanism in quartz based on time-resolved optically stimulated stimulated luminescence *J. Lumin.* **130** 902–9

Chapter 2

Experimental methods

The difference between useful interpretation of experimental data and otherwise is a judicious set of protocols. In this chapter we describe the steps necessary for proper measurement and interpretation of phototransferred thermoluminescence (PTTL). The utility of these procedures is a product of empiricism. Rather than simply listing steps in the protocol for measurement of PTTL, we use test cases of quartz and beryllium oxide (BeO) to illustrate and discuss their need and to highlight the implications of overlooking some of these steps.

2.1 Introduction

If a previously irradiated material, typically an insulator, is heated at a controlled rate, the presence of any electron trapping point defects within it is revealed as intensity peaks in the temperature-resolved plot of luminescence. The electron traps involved in such thermoluminescence (TL) are filled owing to ionization that precedes measurement. If, however, the material is intentionally partially heated to deplete only some of the filled electron traps, and then exposed to light of certain wavelengths to transfer charge from intact more stable (donor) traps to the emptied less stable ones, the resulting PTTL can be informative. Besides providing a means to quantify kinetics of acceptor or donor electron traps, questions about the dynamics of charge transfer and competition effects can be explored. For ease of understanding and to properly set the discussion in context, the basis for the methods to be presented are described with reference to quartz, a widely studied natural mineral and BeO, a refractory binary oxide of long-standing research interest.

The use of the terms donor and acceptor to respectively refer to electron traps from which electrons are optically stimulated and previously emptied ones to which they scatter should be distinguished from the same terminology as used in semiconductor physics. In the latter, the same terms describe the ionization and electron affinity of dopants which are labelled donor-atoms or acceptor-atoms. The terms donor\acceptor electron traps\states are used synonymously throughout this text and in the relevant literature.

doi:10.1088/978-0-7503-3831-8ch2

The literature on conventional thermoluminescence of quartz is extensive. In comparison, the number of studies on PTTL from quartz is meagre. There has been enduring interest in conventional TL of quartz covering such areas as analytical methods, mechanisms and applications as many reviews and texts attest [1–4]. In contrast, except for a few examples that looked at wavelength response [5, 6] or considered emission efficiency (e.g. [7]) or indeed dosimetry [6, 8, 9], the publication of work devoted to PTTL of quartz has been sporadic with many outputs tending to be qualitative.

The possibility of using PTTL for retrospective dosimetry was discussed by Bailiff *et al* [8] on the premise that the amount of residual charge at a donor electron trap reflects the archaeological age. The assumption in this reasoning is that the signal monitored originates from a single donor. Although Bailiff *et al* [8] monitored PTTL at the '110 °C' peak, they also assumed that the donor was the electron trap corresponding to the '325 °C' peak and hence implicitly discounted the possible alternatives at 210 and 375 °C as donors. The study of the possible role of the '325 °C' electron trap as a source of phototransfer goes back a long way [10], has been examined [11–14] and, in some cases, assumed as fact [15]. On the question of PTTL dosimetry, Benny and Bhatt [9] studied quartz with peaks at 110, 220 and 370 °C and yet of these three, the one at 220 °C was not reproduced under phototransfer. This result, although not discussed in their report, hints at competition effects and exemplifies puzzles within PTTL that require systematic investigation.

A number of studies on natural quartz annealed up to 1000 °C [7, 15, 16] or synthetic quartz annealed up to 900 °C [17] also assumed the '325 °C' peak to be the donor for PTTL corresponding to various selected peaks. In what was then an unusual result, Bertucci *et al* [17] observed PTTL in samples preheated to 700 °C. They ascribed the result to deep electron donor traps. A similar conclusion about the role of deep electron traps was drawn by Morris and McKeever [18] and Bailiff *et al* [8]. The presence of deep electron traps in quartz has since been an established fact. In the examples cited above, the dependence of PTTL intensity on the duration of illumination used to induce phototransfer was qualitatively ascribed to donor-to-acceptor charge transfer.

A model based on a linear simple system of one acceptor (the '110 °C' peak) and one donor (the '325 °C' level) was used by Wintle and Murray [14] to describe the PTTL time-response behaviour for quartz. Although the quartz sample used for that work had multiple peaks, the set-up implicitly excluded any contribution from any other electron trap (other than the '325 °C' one) and, for that matter, deep electron traps. A detailed discussion of PTTL intensity profiles was later reported by Alexander and Mckeever [19]. This work also used a system of one donor and one acceptor but was based on numerical simulation and sought to predict expected illumination time-dependent behaviour of PTTL intensity. This study extended earlier ones [5, 14] by including in the model a non-radiative recombination centre and incidences of retrapping thereby rendering the system non-linear. In this approach, the transfer of charge from the donor to the acceptor electron traps during illumination or heating to produce PTTL is formulated as a set of non-linear differential equations. Such coupled equations do not have analytical solutions. By varying some parameters, for example charge concentrations, Alexander and McKeever [19] used numerical solutions to predict some examples of the type of change that PTTL can undergo as a function of the duration of illumination. Similar profiles were shown to result for the same system for various permutations of rates of optical stimulation from electron traps [20]. Thus, patterns obtained by numerical simulation with one set of parameters can also be found when the combination involves a different set of parameters. It must be then that either both or one set is valid or that the exercise is academic.

An enduring aspect concerning the study of PTTL in BeO, our second illustrative case, is the perspective from some of the earliest reports [21, 22]. These studies alluded to the presence of PTTL in samples preheated up to 450 °C. This implied that PTTL is due to phototransfer from deep electron traps and set a paradigm over which the study of PTTL in BeO has been approached. Ever since, measurements intended to examine phototransfer in BeO have routinely been made on samples whose glow curve was depleted of all or most of its peaks. One of the reasons why precedence takes prominence, as in this example, is the lack of systematic protocols for measurement of PTTL.

The notion that analysis built on assumption may not relate to experimental results equally applies to analytical solutions. If one assumes, for a two-trap system, that electrons are removed from the acceptor during illumination and that retrapping is negligible, the solution of the pair of differential equations describing the process is an intensity profile that passes through a maximum with illumination time. If the assumption is invalid, the intensity profile instead increases to saturation. In principle, Alexander and Mckeever [19] cautioned that analytical solutions used to explain the dependence of PTTL on the duration of illumination may or may not be valid for particular systems under experimental study. The system that correctly describes the PTTL from any material under study should not be built on assumptions. These comments also apply for other features of interest such as fading of the PTTL, growth curves or stability of the PTTL.

The analysis of PTTL must be based on experimental evidence. In particular, the number of active donors and acceptors, and to what extent they serve as such, must be informed by experiment. The role of any putative donor should not be presumed but determined by experiment. Supposed acceptors or donors do not always act that way and any causative competition effects need to be investigated and included in any model used to describe the PTTL. We illustrate the procedures for measurement of PTTL by using examples drawn from mostly BeO [23].

2.2 Conventional thermoluminescence: BeO as an exemplar

The first step is the most obvious one—to measure a glow curve and establish its properties. Since the purpose of PTTL measurements is to recover a glow curve or its component peaks by phototransfer, the dynamics and kinetics of the original set of peaks need to be known and understood.

2.2.1 Glow curve features

Figure 2.1 shows glow curves corresponding to 1 Gy and 10 Gy measured from a sample of BeO. The glow curve recorded following 1 Gy shows three nominal peaks at 46, 178 and 290 °C, whereas that for 10 Gy has five at 50, 182, 283, 437 and 535 °C. We denote these peaks as I, II, III, … in that order. In principle, PTTL measurements

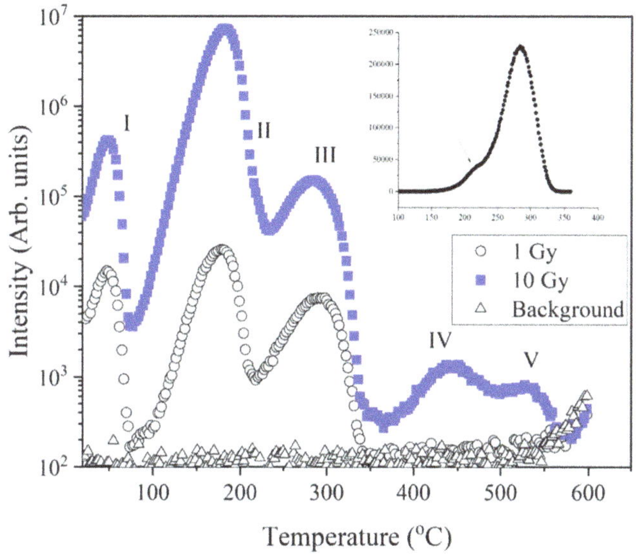

Figure 2.1. Glow curves measured from BeO following irradiation to 1 Gy and 10 Gy. The background signal is included for comparison. There is a weak and improperly defined peak collocated with peak III (inset). Reprinted from [23], with the permission of AIP Publishing.

can be made using any dose but if a signal from deep electron traps is needed, a higher dose is advisable, hence the choice of dose for this case.

2.2.2 Glow curve resolution by thermal cleaning

A glow curve recorded immediately after irradiation properly displays only the prominent peaks. The glow curve therefore needs to be further examined to ascertain its number of peaks since this aids discussion of the PTTL. A useful method for this is the so-called 'thermal cleaning' procedure, which is just shorthand for heating beyond a specific peak to reveal an adjacent one when the full glow curve is recorded. The sample of figure 2.1 was thus heated in turn to 77, 219, 360, 494 and 600 °C after irradiation each time. In addition to the five peaks mentioned earlier, an otherwise obscured peak in the rising edge of peak III appears (figure 2.1, inset). This example is used here to emphasize the point that one should not assume that the prominent peaks in a glow curve are the only ones that compose the glow curve or that only these can be regenerated under phototransfer.

2.2.3 Reproducibility

During the course of PTTL experiments, a sample is likely to be used repeatedly. It is therefore important to assess the possibility that re-use alters the sensitivity of the material to thermal or optical stimulation. In the example of BeO under discussion, the best estimate of peak position was taken as the average and the margin of error as the standard deviation in five consecutive measurements. Peaks I–III were found precisely at 48, 182 and 283 °C and peaks IV and V at 445 ± 3 °C, 534 ± 3 °C. Each peak reaches its maximum within 1.8, 0.7, 1.2, 2.4 and 4.2% of the average. Thus, in this case, all peaks are reproducible.

2.3 Preparatory measurements for PTTL

2.3.1 Active glow peaks

Preparatory tests to determine which glow peaks may appear under phototransfer are essential. It should not be assumed that any or all of the peaks in a glow curve can be reproduced under phototransfer. In our test case, the sample was irradiated and first preheated to 77 °C, illuminated for 10 s and reheated to 600 °C to monitor PTTL. The procedure was repeated by preheating to 219, 360, 494 and 600 °C in turn. This preheating removed peaks I, I–II, I–III, ... before each illumination. Only peaks I–III re-appear under phototransfer. Figure 2.2 shows glow curves corresponding to preheating to 77 °C and 360 °C. A glow curve not preceded by any preheating is included for comparison. Having established which peaks appear under phototransfer, one can then proceed with substantive experiments. In this discussion, we will therefore only be concerned with PTTL peaks I, II and III.

2.3.2 Phototransfer due to 'deeper' electron traps

This section has been included to offer observations on exclusive reliance on precedence and orthodoxy as a guiding step in PTTL measurements. We consider the archetypal perspective on PTTL of BeO, which has been to associate the feature with deep electron traps. Successive studies on the PTTL of BeO have long assumed

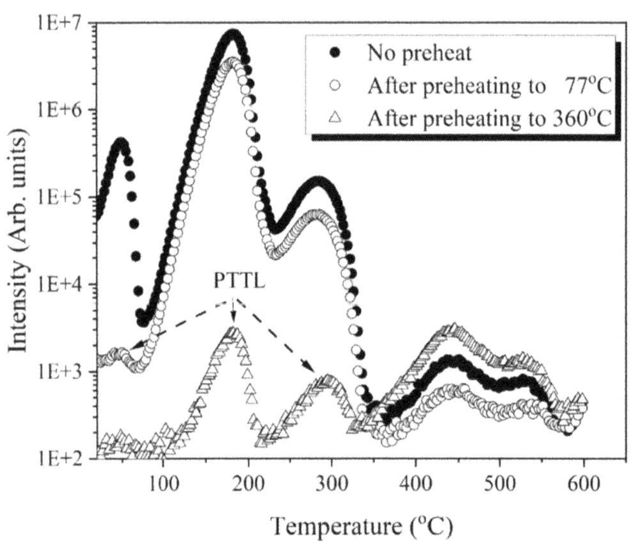

Figure 2.2. Glow curves obtained after preheating to 77 °C and 360 °C and illumination. The original glow curve is shown as are glow curves obtained under phototransfer. Reprinted from [23], with the permission of AIP Publishing.

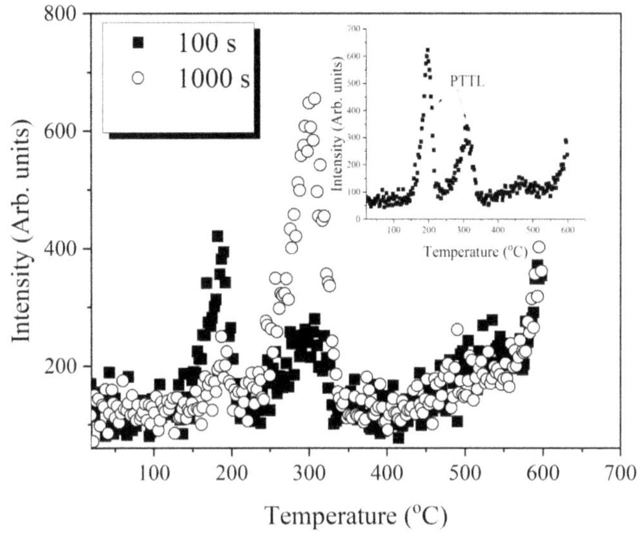

Figure 2.3. In tests intended to examine phototransfer from deep electron traps, PTTL peak II is brighter than PTTL peak III for 100 s illumination. When the duration is increased to 1000 s, the reverse is true. PTTL appears even after annealing at 650 °C for 30 min (inset). Reprinted from [23], with the permission of AIP Publishing.

that the PTTL only originates from deep electron traps. This is akin to the situation in quartz where its PTTL has usually been associated with only the '325 °C' peak. To examine this aspect, a glow curve from the same sample of figure 2.1 was measured following preheating to 500 °C and illumination for 10 s. No PTTL was observed. When the duration of illumination was then lengthened in the follow up measurement, PTTL appeared at peaks II and III. This is shown in figure 2.3. Peak II is more intense for 100 s illumination. When the illumination time is increased to 1000 s, peak III becomes more prominent. There is no PTTL when BeO is preheated to 600 °C. These results imply that donor electron trap(s) for PTTL peaks II and III after preheating to 500 °C lie between 500 °C and 600 °C. It may then be assumed that the electron trap for peak V contributes to PTTL seen at peaks I–III. Since only lengthy illumination is required to induce this PTTL, the contribution from the said donor is minor. The tests did not show any evidence of electron traps 'deeper than 600 °C' contributing to the PTTL. It has been reported [22, 24] that heating BeO to 650 °C 'removes the PTTL effect'. It is open to interpretation whether this means that such heating extirpates capacity for PTTL or that PTTL cannot be observed following preheating to 650 °C but is present otherwise. The latter is consistent with our findings as explained earlier. To examine the possibility of extirpation, measurements were made on a sample annealed at 650 °C for 30 min. When the sample was irradiated to 50 Gy and preheated to 77, 219 and 360 °C, PTTL was observed at peaks II and III (figure 2.3, inset). One concludes then that PTTL is produced from BeO despite annealing at 650 °C.

2.3.3 Fading

2.3.3.1 Light-induced fading

The optically-induced fading of glow peaks may indicate which of them may be involved in the PTTL. This detail, although only supplementary, may assist in interpretation of the PTTL. To study this, the residual TL can be monitored in turn from an irradiated sample illuminated for certain periods before any heating. We illustrate this on the BeO in question. Owing to differences in intensity of the peaks, glow curves were measured for two doses, one a hundred times greater than the other. To monitor peaks I–III, 0.1 Gy was used whereas for peaks IV and V the sample was irradiated to 10 Gy. All peaks fade with illumination (figure 2.4(a)). Thus, electron traps of all peaks should be involved in the PTTL as acceptors,

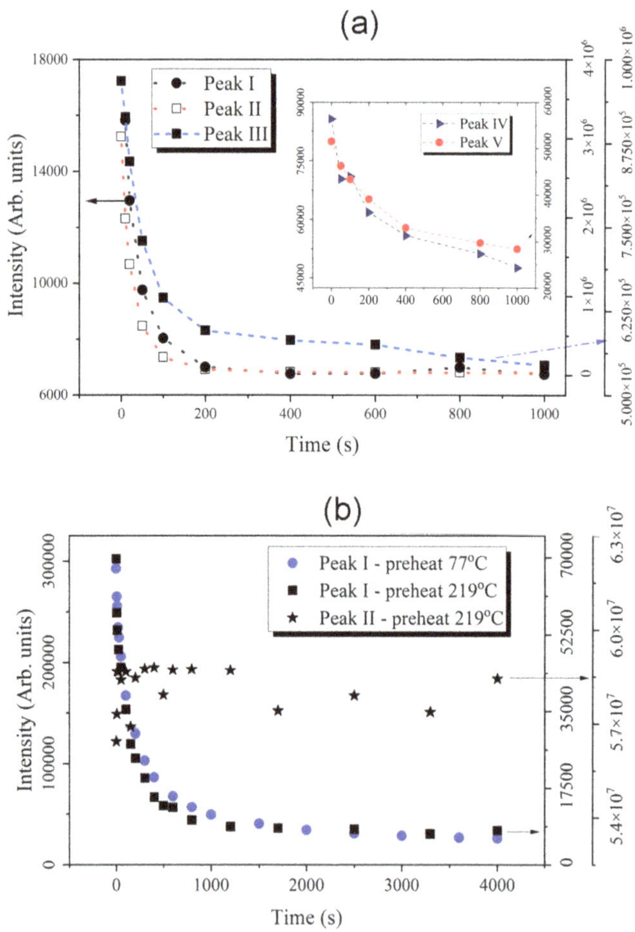

Figure 2.4. The effect of delay between irradiation and measurement for peaks 1–V corresponding to various preheating as shown. The inset to part (b) is for peak II following preheating to 219 °C. Optically stimulated luminescence could be measured in each case. This example is for measurements following preheating to 77 °C and 219 °C. Reprinted from [23], with the permission of AIP Publishing.

donors or competitors. This also demonstrates that electron traps other than deep ones are involved in the PTTL and counters the common assumption that PTTL only ensues from deep electron traps. The question of whether and how the electron trap of any peak that is not optically induced to fade may affect PTTL, is as yet, open.

2.3.3.2 Thermal fading of PTTL

Figure 2.4(b) shows intensity against the time after irradiation for PTTL peaks I and II for preheating to 77 °C and 219 °C intended to remove peak I and peaks I-II, respectively. Peak I fades between irradiation and measurement. In this time, phosphorescence is emitted showing that the fading is thermally induced. In comparison, the intensity of PTTL peak II is not affected by the delay between irradiation and measurement.

2.4 Identification of donor and acceptor electron traps by pulse annealing

Although it is often assumed that all electron traps corresponding to peaks in a glow curve may influence phototansfer as acceptors or donors, it is necessary to determine the extent to which they do. In the example of BeO, the PTTL intensity corresponding to illumination for 10 s was monitored each time a sample irradiated to 10 Gy was preheated. Preheating temperatures were increased from 80 °C to 650 °C at 10 °C intervals. The intensity of peaks not removed by preheating was also noted. The peaks labels mentioned below refer to figure 2.1.

2.4.1 Active glow peaks

Figure 2.5 shows the influence of preheating temperature on the intensity of peaks I–V. To start with, the intensity of each of the peaks I and II decreases with preheating (figure 2.5(a)). The change for peak II is slow up to 180 °C but picks up thereafter until 200 °C as the peak itself is removed by the preheating. This is accompanied by a decrease in the intensity of the PTTL observed at peak I. This suggests that the electron trap for peak II acts as a donor for that phototransfer. Peaks I and II then become stable in intensity between 180 °C and 280 °C, presumably because their donor is unaffected by preheating in this temperature range. Increasing the preheating temperature to 320 °C causes the intensities of PTTL peaks I and II to drop considerably. The decrease of PTTL seen at peaks I and II occurs simultaneously with the lowering in intensity of peak III. One concludes then that the electron trap for peak III is a donor for PTTL monitored at peak II and an additional donor for PTTL observed at peak I. When preheating extends beyond 320 °C, PTTL only appears at peak II as long as peak IV is present (figure 2.5(a), inset). This suggests that the contribution of the electron trap for peak V to the PTTL at peak II is negligible. This is the same conclusion drawn in our earlier discussion about PTTL from deep electron traps (section 2.3.2). PTTL peak II is therefore due to phototransfer from electron traps for peaks III and IV, with that from peak III being more important.

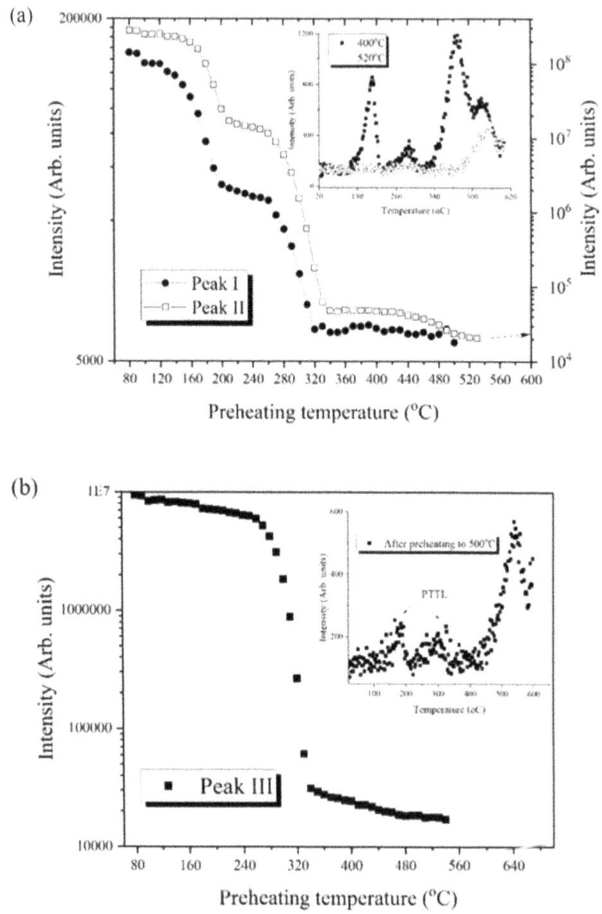

Figure 2.5. Change of intensity with preheating temperature for peaks I and II (a), III (b). The inset to part (a) shows measurements following preheating to 400 °C and 500 °C. Reprinted from [23], with the permission of AIP Publishing.

Figure 2.5(b) singles out the influence of preheating temperature on the intensity of peak III. Thc intensity of this peak weakens between 80 °C and 260 °C because it serves as a donor for PTTL peaks I and II between these temperatures. The subsequent rapid decrease in its intensity up to 340 °C occurs because the peak is progressively removed by preheating. Peak III thereafter reappears under phototransfer, which incidentally, offers a means to assess the role of its donors. When the contribution of its supposed donor electron traps, namely, the electron traps for peaks IV and V (figure 2.6) decreases, so does the intensity of peak III beyond 340 °C. We note that peak III still appears even after peak IV near 500 °C has been removed. Figure 2.5((b), inset) shows an example of a glow curve measured after preheating to 500 °C with only peak V remaining. The presence of PTTL peaks II and III here is obvious.

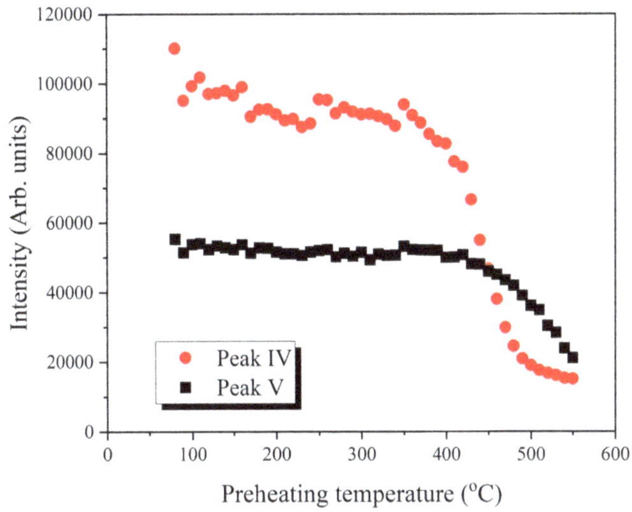

Figure 2.6. The dependence of intensity on preheating temperature for peaks IV and V. Reprinted from [23], with the permission of AIP Publishing.

2.4.2 Quiescent glow peaks

Peaks IV and V are not regenerated by phototransfer. These peaks are not affected by preheating other than being removed by the preheating itself (figure 2.6). This shows that their electron traps are only weak donors for any PTTL at peaks I–III.

2.5 Key steps for PTTL measurement

The main experimental steps in the measurement of PTTL may be listed as follows:

1. The first step is the obvious one of measuring a glow curve. Since as many peaks as the instrumentation allows need to be monitored, the choice of final temperature should not be conservative.
2. It is inevitable that a sample will be heated multiple times during PTTL experiments. Such re-use can change the sensitivity of the sample and perhaps also cause changes to peak positions. At the outset, the glow curve should therefore be measured several times to ascertain that the position and intensity of peaks are stable for a given dose. The effect of any instability in these parameters on the PTTL cannot be ignored.
3. In most cases, it is possible to make out the number of peaks in a glow curve. Nevertheless, their presence should be systematically verified by partial heating procedures such as thermal cleaning or the T_m-T_{stop} method. Establishing the complete number of peaks and their position provides some confidence when deciding the number of peaks associated with donor electron traps. This step also helps to reduce reliance on guesswork, which has been the bane of some studies on PTTL.

4. Since phototransfer optically decreases the concentration of trapped electrons, it is advisable to check which peaks are affected if the sample is illuminated with the same light that is to be used in PTTL experiments. If a peak is involved in phototransfer, its intensity should decrease with illumination. Whether this applies to peaks whose intensities are unaffected is a cause for investigation.
5. To determine the extent to which electron traps influence the phototransfer as acceptors or donors, the PTTL intensity corresponding to illumination for a fixed period ought to be monitored each time a sample irradiated to a given dose is preheated. Preheating temperatures should cover the expanse of the glow curve. The intensity of all peaks including any not removed by preheating should be monitored.
6. Lastly, it is essential to investigate the expectation that PTTL will be produced at a given peak once selected ones have been removed by preheating. Once the sample has been irradiated, it should be illuminated after preheating to several temperatures chosen to remove each peak in turn. This is one way to have a qualitative idea of which peak(s) may be associated with a donor. The appearance of PTTL after only one peak has been removed is only a fortunate exception.

The strategies used to study phototransfer are illustrated in figure 2.7. The main idea is that an irradiated and preheated sample is illuminated to transfer electrons from deeper to shallower electron traps and then reheated to monitor any PTTL peaks

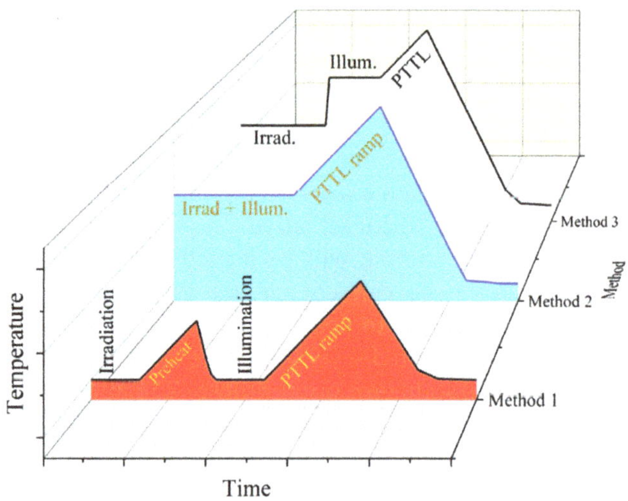

Figure 2.7. Schematic representation of strategies used to study phototransfer in this work. An irradiated and preheated sample is illuminated to transfer electrons from deeper to shallower electron traps and then reheated to monitor any PTTL peaks below the preheating temperature (method 1). If the irradiation or illumination temperature is identical to or exceeds the preheating temperature, preheating becomes redundant (methods 2 and 3). Reprinted from [25], with the permission of AIP Publishing.

below the preheating temperature. This is method 1. If the irradiation or illumination temperature is identical to or above the preheating temperature, preheating becomes redundant (methods 2 and 3). The procedures described in this chapter mostly used method 1 for discussion but equally apply to methods 2 and 3 with suitable modifications.

References

[1] Preusser F, Chithambo M L, Götte T, Martini M, Ramseyer K, Sendezera E J, Susino G J and Wintle A G 2009 Quartz as a natural luminescence dosimeter *Earth Sci. Rev.* **97** 196–226

[2] Bøtter-Jensen L, McKeever S W S and Wintle A G 2003 *Optically Stimulated Luminescence Dosimetry* (Amsterdam: Elsevier)

[3] McKeever S W S 1985 *Thermoluminescence of Solids* (Cambridge: Cambridge University Press)

[4] Chen R and McKeever S W S 1997 *Theory of Thermoluminescence and Related Phenomena* (Singapore: World Scientific)

[5] Alexander C S, Morris M F and McKeever S W S 1997 The time and wavelength response of phototransferred thermoluminescence in natural and synthetic quartz *Radiat. Meas.* **27** 153–9

[6] Colyott L E, Akselrod M S and McKeever S W S 1996 Phototransferred thermoluminescence in Alpha Al_2O_3:C *Radiat. Prot. Dosim.* **65** 263–6

[7] Milanovich-Reichhalter I and Vana N 1990 Phototransferred thermoluminescence in quartz *Radiat. Prot. Dosim.* **33** 211–3

[8] Bailiff I K, Bowman S G E, Mobbs S F and Aitken M J 1977 The phototransfer technique and its use in thermoluminescence dating *J. Electr.* **3** 269–80

[9] Benny P G and Bhatt B C 2002 High-level gamma dosimetry using phototransferred thermoluminescence in quartz. *Appl. Radiat. Isotopes.* **56** 891–4

[10] Schlesinger M 1965 Optical studies of electron and hole trapping levels in quartz *J. Phys. Chem. Solids* **26** 1761

[11] Aitken M J 1998 *An Introduction to Optical Dating* (Oxford: Oxford University Press)

[12] Spooner N A 1994 On the optical dating signal from quartz *Radiat. Meas.* **23** 593–600

[13] Smith B W, Aitken M J, Rhodes E J, Robinson P D and Geldard D M 1986 Optical dating: methodological aspects *Radiat. Prot. Dosim.* **17** 229–33

[14] Wintle A G and Murray A S 1997 The relationship between quartz thermoluminescence, photo-transferred thermoluminescence, and optically stimulated luminescence *Radiat. Meas.* **27** 611–24

[15] Santos A J J, de Lima J F and Valerio M E G 2001 Luminescent, optical and color properties of natural rose quartz *Radiat. Meas.* **33** 427–30

[16] Milanovich-Reichhalter I and Vana N 1991 Phototransferred thermoluminescence in quartz annealed at 1000 °C *Nucl. Tracks Radiat. Meas.* **18** 67–9

[17] Bertucci M, Veronese I and Cantone M C 2011 Photo-transferred thermoluminescence from deep traps in quartz *Radiat. Meas.* **46** 588–90

[18] Morris M F and McKeever S W S 1994 Optical bleaching studies of quartz *Radiat. Meas.* **23** 323–7

[19] Alexander C S and McKeever S W S 1998 Phototransferred thermoluminescence *J. Phys. Appl. Phys.* **31** 2908

[20] Moscovitch M 2011 The principles of phototransferred thermoluminescence *AIP Conf. Proc.* **1345** 323–34

[21] Tochlin E, Goldstein N and Miller W G 1969 *Health Phys.* **16** 1
[22] Crase K W and Gammage R B 1975 *Health Phys.* **29** 739
[23] Chithambo M L and Kalita J M 2021 *J. Appl. Phys.* **130** 195101
[24] Bulur E and Göksu H Y 1998 *Radiat. Meas.* **29** 639
[25] Chithambo M L 2022 Phototransferred thermoluminescence of CaF_2: principles, analytical methods and mechanisms *J. Appl. Phys.* **132** 055102

IOP Publishing

Phototransferred Thermoluminescence

Makaiko L Chithambo

Chapter 3

Analytical methods

3.1 Introduction

Phototransferred thermoluminescence *is* thermoluminescence. The mechanisms, models and applications of thermoluminescence are well documented (e.g. [1–3]). Practical aspects of its analysis are outlined elsewhere (e.g. [4–6]). PTTL is distinguished from conventional TL in how each is generated. Whereas conventional TL ensues due to the filling of electron traps owing to ionizing radiation, PTTL is a restorative effect caused by light transferring electrons from some of the pre-filled electron traps to purposely emptied ones. The experimental steps for measurement of PTTL are usually as follows:

(a) The sample is irradiated during which electrons move to certain imperfections serving as electron traps.
(b) The material is then preheated to deplete shallower electron traps.
(c) This is followed by illumination to induce phototransfer.
(d) Finally, the sample is heated at a controlled rate to record a glow curve during which PTTL is monitored.

The experimental protocols need not include all steps listed. If the sample is irradiated at the same temperature it is preheated to, preheating becomes redundant and step (b) can be omitted. The method also assumes that illumination only empties electron traps and does not cause ionization as the irradiation in step (a) does. Successful phototransfer is indicated by the reappearance of any glow peak that was preheated out.

PTTL was first reported by Stoddard [7] for measurements on sodium chloride. That work and subsequent studies set a paradigm over which PTTL was studied. Although use of UV illumination to achieve phototransfer is a deeply rooted practice, it is not requisite for phototransfer to occur. Depending on the material under study, light of various wavelengths spanning UV to the infrared can be used to induce phototransfer. It is only their capability for phototransfer that differs.

doi:10.1088/978-0-7503-3831-8ch3

For some purposes, samples can be irradiated at room temperature and then cooled down to cryogenic temperatures at which they are illuminated [8, 9]. In that way, electron traps that are too unstable to retain charge when the material is irradiated at room or higher temperatures, can be sampled. This chapter, however, is not concerned with this method.

If phototransfer occurs in line with theoretical expectations, the appearance of a PTTL peak should be accompanied by a decrease in the intensity of some, if not all, peaks left intact during preheating. Experimentally, this does not always occur and this reality is usually overlooked because in most cases the only intensity monitored is that of the acceptor peak. Indeed, the intensity of supposed donor peaks sometimes counterintuitively increases or remains independent of illumination. With this and other examples, PTTL offers a rich tapestry of problems and puzzles for analytical study. In particular, the analysis of the dependence of PTTL intensity on the duration of illumination remains an enduring concern. Besides this, the dependence of the PTTL intensity on the combination of dose, preheating temperature and illumination time has not been studied to any significant extent. Other mostly open questions include the role of illumination wavelength on the process, the effect of change in importance of donors during measurement, competition between acceptor and donor electron traps, the effect of illumination and irradiation temperature on the PTTL as well as the effect of hole traps and deep electron traps on PTTL.

The aim of this chapter is twofold. We first discuss the analysis of time-response profiles, i.e. the dependence of PTTL intensity on the duration of illumination, and then look at the influence of illumination temperature on PTTL. The analytical methods discussed are the kinetic and phenomenological models. Because of their utility, the complementary techniques of vector fields and computational simulation are included.

3.2 Kinetics model

The first substantive study of kinetic models of PTTL was reported by Alexander and McKeever [10]. Their exemplar was a case of two electron traps and one recombination centre, the same one used by Wintle and Murray [11] to discuss PTTL in natural quartz. The models of Alexander and McKeever [10] sought to explain the dependence of PTTL intensity on the duration of illumination. The study also considered a case of two electron traps and two recombination centres, one radiative and the other, non-radiative.

In the kinetic description of PTTL, rate equations relate the concentration of holes at a recombination centre to the concentration of electrons at electron traps and in the conduction band. The resulting set of coupled differential equations is non-linear and as such does not have an analytical solution and requires a numerical approximation to one. If an analytical solution is needed, one invokes simplifying assumptions to derive it.

3.2.1 Two electron traps and one recombination centre

The model discussed by Alexander and McKeever [10] consists of two electron traps and one recombination centre. This is a system of one acceptor and one donor. Electron transfer between the electron traps is described to occur through the conduction band. When the sample is preheated after irradiation, shallow electron trap(s) are cleared. By the charge neutrality condition, the concentrations of holes in the recombination centre and of electrons in donor electron traps must balance. The charge transport during illumination can be expressed as

$$\frac{dn_a}{dt} = -f_a n_a + n_c(N_a - n_a)A_a \tag{3.1}$$

$$\frac{dn_d}{dt} = n_c(N_d - n_d)A_d - f_d n_d \tag{3.2}$$

$$\frac{dn_c}{dt} = -\frac{dn_a}{dt} - \frac{dn_d}{dt} + \frac{dm}{dt} \tag{3.3}$$

$$\frac{dm}{dt} = -A_m m n_c \tag{3.4}$$

where N_d and N_a denote the concentration of donor and acceptor electron traps whose respective instantaneous electron concentrations are n_{di} and n_a. The parameters A_d and A_a are the retrapping probabilities and A_m the recombination probability; n_c and m are the instantaneous concentration of electrons in the conduction band and holes at the recombination centre and f_d and f_a are the optical stimulation probabilities.

The set of rate equations that describe the measurement of the PTTL peak when the material is heated after illumination is

$$\frac{dn_d}{dt} = n_c(N_d - n_d)A_d \tag{3.5}$$

$$\frac{dn_a}{dt} = -p_a n_a + n_c(N_a - n_a)A_a \tag{3.6}$$

$$\frac{dn_c}{dt} = -\frac{dn_a}{dt} - \frac{dn_d}{dt} + \frac{dm}{dt} \tag{3.7}$$

$$\frac{dm}{dt} = -A_m m n_c \tag{3.8}$$

where p_a is the thermal detrapping rate given as

$$p_a = s\exp(-E_a/kT) \tag{3.9}$$

where k is Boltzmann's constant, T is temperature, E_a is the activation energy of the acceptor electron trap and the so-called frequency factor s is a measure of the number of times a trapped electron attempts to detach from its binding potential [11]. This set consists of non-linear differential equations and does not have an analytical solution. One way to derive one is to invoke a number of assumptions. For example, assuming the quasi-equilibrium approximation, namely,

$$\left|\frac{dn_c}{dt}\right| \ll \left|\frac{dm}{dt}\right| \tag{3.10}$$

leads to

$$I_{\text{PTTL}} = -\frac{dm}{dt} = p_a f(m) \tag{3.11}$$

where I_{PTTL} is the intensity of the one PTTL peak monitored. The exact form of the function $f(m)$ is not essential for the discussion here but is discussed elsewhere [2]. As pointed out [9], the numerical solution obtained will be descriptive only for parameters in the expression obtained.

The measurement of a glow curve during which the resulting PTTL peak is picked out is normally made as a function of temperature. Using the heating rate $\beta = dT/dt$ as a lead to change the independent variable and assuming backscatter into the donor electron trap, it can be shown [2, 9] that the area A_{PTTL}under a PTTL peak can be written as

$$A_{\text{PTTL}} \approx \frac{A_m m_i n_{ai}}{A_d(N_d - n_{di})} \tag{3.12}$$

if $n_{ai} \ll N_d - n_{di}$ where n_{ai} and n_{di} denote initial values. Unless the assumption for retrapping into the donor electron trap is dropped, equation (3.12) suggests that the PTTL intensity will not only be proportional to the initial concentration of electrons at the acceptor electron trap but will also be a factor of the other initial concentrations as shown. This implies that if equation (3.12) is to be used to analyse the dependence of PTTL intensity on duration of illumination, the time dependencies $n_{ai}(t)$, $n_{di}(t)$ and $m_i(t)$ need to be known.

When retrapping into the donor trap or optically induced loss from the acceptor electron trap no longer apply occur, it follows that

$$A_{\text{PTTL}} \propto \left[1 - \exp\left(-f_d t\right)\right] \tag{3.13}$$

This predicts that the PTTL intensity will increase to saturation with illumination time.

Confining the discussion to the optical stimulation stage only as in equations (3.1)–(3.4) and assuming no retrapping into the source electron trap, the time dependence of the PTTL intensity for a system of one donor and one acceptor is found as

$$A_{\text{PTTL}} = C\left[\exp\left(-f_d t\right) - \exp\left(-f_a t\right)\right] \tag{3.14}$$

where $C = \delta f_d n_{di}/(f_a - f_d)$ and δ a constant of proportionality. This function is a difference of exponentials and increases through a maximum with duration of illumination. If there is no loss of electrons from the acceptor electron trap during illumination, the time-dependent change of the intensity reduces once more to a saturating exponential, namely,

$$A_{\text{PTTL}} = Cn_{di}[1 - \exp\left(-f_d t\right)] \tag{3.15}$$

This expression and the preceding ones refer to a system of one acceptor and one donor and have been obtained using certain assumptions. The last example has also been expounded on by Furetta [12]. The equations arrived at in these works [2, 12] are not generic and apply to the cases discussed under the assumptions made. Indeed, the PTTL of most materials can be understood in terms of one acceptor and multiple donors. The tutorial model of one acceptor and one donor is an experimental exception. The profile of a saturating intensity with illumination time as discussed can also be obtained for the same system for various combinations of rates of optical stimulation from electron traps [9, 13]. Thus, results obtained with one set of parameters can equally be found when the permutation involves a completely different set of values for the same parameters. For the exercise to be other than academic, assumptions must be justified and where possible, experimentally verified.

The need to be circumspect about assumptions also applies to analytical solutions because corresponding results may not relate to the experimental outcome. If one supposes, for a two electron trap system, that there is loss of charge from the acceptor during illumination, the solution of the defining set of rate equations is an intensity profile that passes through a peak with illumination time. When the assumption is dropped, the intensity change resembles a saturating exponential. For PTTL emission over lengthy time-scales as is usual, use of analytical functions is meaningful but becomes less so when the emission is transient or does not have any obvious dependence on illumination time. Thus, analytical solutions derived to account for time dependence of PTTL may or may not be valid for particular systems under experimental study.

3.3 Phenomenological model

3.3.1 Compartmental analysis

A useful device to help formulate the transport of electrons between systems of acceptor and donor electron traps under PTTL that is relevant for the phenomenological model is compartmental analysis. The aim of compartmental analysis as discussed in various texts (e.g. [14]) is to segment a system into a set of subsystems or compartments. Each compartment can be described by one differential equation but since the compartments interact by exchanging particles, the whole system can be described by a system of coupled differential equations. The exchange of particles

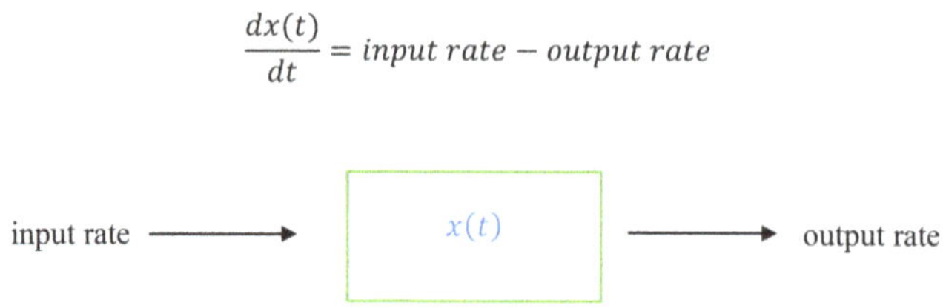

Figure 3.1. A particle transport diagram for a single compartmental system.

may adhere to certain constraints of the system under study and this is taken into account in the mathematical description of the system.

If the parameter $x(t)$ represents, say the concentration of itinerant particles, the rate of change of concentration of particles in a single compartment such as that shown in figure 3.1 is

$$\frac{dx(t)}{dt} = \text{input rate}-\text{output rate}$$

Since the input and output rates can be defined in terms of parameters of the whole system, intermediate stages can be omitted. The number of compartments or stages used to generate the set of rate equations will depend on the problem at hand. In particular, in applying the concept to PTTL, it is advisable to analyse the system on the basis of verifiable assumptions.

3.3.2 System of one acceptor multiple donors

We treat the emission of PTTL as corresponding to a system of one acceptor, labelled say k, and n donors whose instantaneous concentration of trapped electrons is N_i, where the index i enumerates donors. It is assumed that during illumination electrons are optically stimulated from a donor at a rate $f = \Phi\sigma$, where Φ is the incident photon flux and σ the photoionization cross-section. Only some of the electrons removed from a specific donor to the conduction band transit to an acceptor electron trap. This is to be expected since electrons moved to the conduction band from donors can also move to recombination centres or scatter back to donor electron traps. In describing the influence of duration of illumination on PTTL intensity, we are only concerned with the portion that ends up at acceptors. If the number of electrons optically removed from the ith donor is $f_i N_i$, the number that scatters to an acceptor is $\delta_i f_i N_i$, where δ_i is a proportionality constant. Whether or not retrapping is relevant should ideally be decided by experiment. The thermoluminescence intensity is taken to be proportional to the time-dependent concentration of electrons at the acceptor. This is not strictly accurate as the intensity is also affected by the initial concentrations at the donor and recombination centres but is nevertheless a good approximation.

The phototransfer of electrons from donors to an acceptor can be expressed more compactly in matrix form as

$$N' = AN \tag{3.16}$$

where N are eigenvectors and A is the matrix of coefficients. The basic assumption here is that any peak not removed by preheating may correspond to a donor. This assumption can be experimentally investigated. Supposed donors whose contribution turns out to be immaterial can be ignored in the analysis.

Under conditions of negligible retrapping and where there is some optical stimulation from the acceptor, the transport of electrons from n donors to an acceptor can be expressed as

$$N' = \begin{bmatrix} -f_n & \cdots & 0 \\ \vdots & \ddots & \vdots \\ \alpha_n f_n & \cdots & -f_k \end{bmatrix} \begin{bmatrix} N_n \\ \vdots \\ N_k \end{bmatrix} \tag{3.17}$$

where all symbols have their previous meanings. Using the characteristic equation of A, namely, $| A - m\boldsymbol{I} | = 0$, where m are eigenvalues of A and $\boldsymbol{I}$ is a unit matrix, it follows that $= -f_j$ and $-f_k$ where the terms f_j relate to acceptors. The initial value n_{ai} always has a finite value and as such equation (3.17) will have a particular solution giving the time dependence of the PTTL for a system of one acceptor and n donors expressed as

$$N_k = \sum_{j=1}^{n} C_j \; (e^{-f_j t} - e^{f_k t}) \tag{3.18}$$

where $C_j = \alpha_{jj} f_j N_{ij} / (f_k - f_j)$. For reference, some background on how to use matrices to solve linear systems is given in box 3.1.

Box 3.1. Use of matrices to solve linear systems.
Consider a linear and homogeneous system of ordinary differential equations with constant coefficients. For convenience, we discuss a 2 × 2 system of form

$$x' = ax + bz \tag{3.19}$$

$$z' = kx + lz \tag{3.20}$$

where a, b, k and l are constants. This system can be rewritten in matrix form as

$$\begin{pmatrix} x' \\ z' \end{pmatrix} = \begin{pmatrix} a & b \\ k & l \end{pmatrix} \begin{pmatrix} x \\ z \end{pmatrix} \tag{3.21}$$

We proceed to look for solutions of type

$$\begin{pmatrix} x \\ z \end{pmatrix} = \begin{pmatrix} \gamma_1 \\ \gamma_2 \end{pmatrix} e^{mt} = \begin{pmatrix} \gamma_1 e^{mt} \\ \gamma_2 e^{mt} \end{pmatrix} \tag{3.22}$$

where γ_1, γ_2 and m are uknown and yet to be determined constants. To find the constants, we substitute equation (3.22) into equation (3.21), hence

$$m \begin{pmatrix} \gamma_1 \\ \gamma_2 \end{pmatrix} e^{mt} = \begin{pmatrix} a & b \\ k & l \end{pmatrix} \begin{pmatrix} \gamma_1 \\ \gamma_2 \end{pmatrix} e^{mt} \tag{3.23}$$

from which

$$m\begin{pmatrix}\gamma_1\\ \gamma_2\end{pmatrix} = \begin{pmatrix}a & b\\ k & l\end{pmatrix}\begin{pmatrix}\gamma_1\\ \gamma_2\end{pmatrix} \tag{3.24}$$

Equation (3.24) cannot be simplified by simple subtraction since the scalar m on the left-hand side cannot be subtracted from the leading matrix on the right-hand side. To resolve this, the scalar m is replaced by the diagonal matrix $m\boldsymbol{I}$ where $\boldsymbol{I}$ is a unit matrix

Equation (3.24) is a pair of non-linear equations in three variables and in that sense is intractable. If we instead regard the equations as a pair of linear equations in the unknowns a, b, k and l, then the equations can be expressed in standard form as

$$(a - m)\gamma_1 + b\gamma_2 = 0 \tag{3.25}$$

$$k\gamma_1 + (l - m)\gamma_2 = 0 \tag{3.26}$$

Equations (3.25) and (3.26) are then a pair of homogeneous linear equations. These types of equation have a non-zero solution *if and only if* the determinant of coefficients is equal to zero, hence

$$\begin{vmatrix} a - m & b \\ k & l - m \end{vmatrix} = 0 \tag{3.27}$$

from which

$$m^2 - (a + l)m + (al - bk) = 0 \tag{3.28}$$

Equation (3.28) can be solved for m. Once found, values of m can be substituted into (3.25) and (3.26) to obtain γ_1 and γ_2. Values of m, γ_1 and γ_2 can then be substituted into equation (3.22) to generate solutions to the system of equations (3.19) and (3.20). Equation (3.28) is the **characteristic equation** of the matrix of coefficients

$$A = \begin{pmatrix}a & b\\ k & l\end{pmatrix} \tag{3.29}$$

The **eigenvalues** are the roots m_i of the characteristic equation.

The set of equations (3.19) and (3.20) can be written as

$$\boldsymbol{x}' = A\boldsymbol{x} \tag{3.30}$$

where A is the matrix of coefficients and $\boldsymbol{x}'$ and $\boldsymbol{x}$ are **vectors**. Therefore, the characteristic equation can rewritten as

$$| A - m\boldsymbol{I} | = 0 \tag{3.31}$$

where $\boldsymbol{I}$ is the corresponding unit matrix.

The characteristic equation for an nxn matrix, namely, is

$$| A - m\boldsymbol{I} | = (-m)^n + tr(A)(-m)^{n-1} + \ldots + \det A = 0 \tag{3.32}$$

where $tr(A)$ is the trace, that is, the sum of diagonal elements and det A, the **determinant**. This characteristic polynomial is of degree n and (3.32) has at most n real eigenvalues. The corresponding determinant is

$$\det = \sum(A_{ij})\left[(-1)^{i+j}M_{ij}\right] \tag{3.33}$$

where A_{ij} denotes an element in row i and column j of the matrix. The parameter M_{ij} is the minor of A_{ij} and is found by working out the determinant of the matrix that remains after crossing off the row and column containing the element A_{ij}.

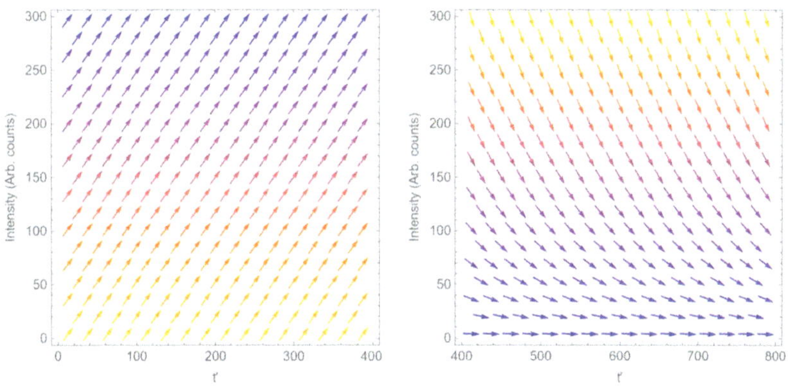

Figure 3.2. Vector fields of analogues of expressions relating to measurement of PTTL for certain electron traps in tanzanite [15]. The plots display the general shape of their solutions. Reprinted from [15], Copyright (2021), with permission from Elsevier.

3.4 Vector fields

Analytical solutions discussed thus far are approximations derived with the use of several assumptions. This is inevitable because other factors such as the effect of preheating in reducing the concentration at electron traps cannot always be modelled precisely. To complement the kinetic and phenomenological models, vector fields can be instructive. Vector fields display the general shape of solutions of differential equations between chosen boundaries. The aim of a vector field is to provide a way to trace the trajectory of a solution in the phase plane, that is, the 'xy' plane. In doing so, it is not necessary to know the solution of the equation or set of equations in question. The inclination of the vector fields and trajectories that can be traced in the phase plane can be compared with experimental results. Figure 3.2 shows examples of vector field analogues of equations relating to PTTL from tanzanite [15]. The important point here is that the fields offer an idea of the trajectory of any unique solution.

3.5 Simulation

Most systems of equations describing the electron transport from donors to an acceptor cannot be solved analytically. As an alternative to vector fields, one can set initial conditions, say for values of n_{ai}, n_{di}, m, etc, that relate to experiment and in that way obtain and graph a unique numerical solution which can be compared to experimental behaviour. Various examples of this approach have been reported elsewhere (e.g. [16]). A detailed account of the use of simulation in thermally and optically stimulated luminescence has been given by Chen and Pagonis [17].

3.6 Stability

The limiting behaviour of the systems under small deviations in the initial conditions can be studied by analysing the stability of the solutions. The basis of this method goes back to the same reason that systems of equation as discussed are in many instances non-linear and hence intractable. We therefore linearize them in the neighbourhood of a critical point by using the procedure

$$\vec{U'} = \vec{J}\vec{U} \tag{3.34}$$

where J is the Jacobian and $\vec{U}$ a suitable vector that relates to the problem in question. Since the system will have a solution only if $(\vec{J} - mI)$ vanishes, the resulting eigenvalues m are substituted into the Jacobian and predict various types of equilibrium solutions. The final step is to attempt to interpret the physical meaning of the critical points in the phase portrait. An example of the application of stability theory for PTTL was discussed for CaF_2 in reference [18]. The study showed that the concentrations of holes and electrons cannot mutually annihilate and there will always be an equilibrium value for both that will asymptotically approach but not actually reach zero.

3.7 Quantifying the role of donor electron traps

The contribution of putative donors can be quantified using the intensity of their corresponding peaks for each preheating. This can be done by way of ratios of the most prominent peak to all others in the set of donors corresponding to a particular acceptor. An example of such a plot is shown in figure 3.3 for measurements made from BeO [16]. The preheating temperatures shown were part of the experiment used in that study. Figure 3.3 shows that although PTTL measured after preheating to 77 °C (intended to remove the first peak) corresponds to four putative donors, only two make a material contribution to the process, hence its system can be analysed as one of two donors. The PTTL after preheating to 219 °C has three donors whereas that for preheating to 360 °C has two. In this way, the analysis of the PTTL can be informed by experiment and not be wholly reliant on assumption.

In some cases, it may be impractical to monitor and quantify the role of a particular donor. This would be the case, for example, for deep electron traps that activate beyond the heating limit of the measurement apparatus. In such cases, it is helpful to mathematically verify the qualitative interpretation of experimental data. A relevant example is discussed for PTTL of α-Al_2O_3:C [19] as monitored at the shallow trap (labelled as peak 1). The PTTL is associated with two donors (peaks 3 and 4 as labelled in that study). The rate of electron transfer during the illumination assuming no retrapping into the donor is

$$\frac{dN_3}{dt} = -f_3 N_3 \tag{3.35}$$

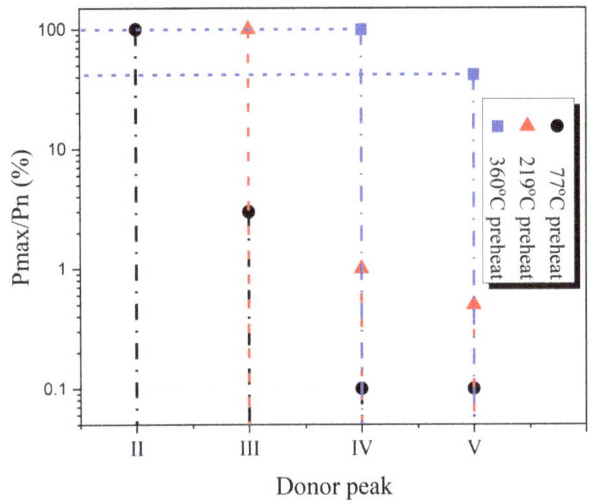

Figure 3.3. A plot of ratios of the dominant donor to all others in each set corresponding to a given preheating temperature in measurements of PTTL from BeO. Reprinted from [16], with the permission of AIP Publishing. Note that the label for the ordinate, when referred to the calculations, is P_n/P_{max} and not as reversed in [16], here and elsewhere in this book.

$$\frac{dN_4}{dt} = -f_4 N_4 \tag{3.36}$$

$$\frac{dN_1}{dt} = -f_1 N_1 + \alpha f_3 N_3 + b f_4 N_4 \tag{3.37}$$

where α and b are proportionality constants. The donor with label 4 is the deep electron trap. In practice it is observed that for the preheating involved, the deep trap makes a minor contribution to phototransfer. For this reason, one deduces that the influence of the third term in equation (3.37) is negligible and can be written $h(t) = \gamma f_4 N_4$, hence

$$\frac{dN_1}{dt} = -f_1 N_1 + \alpha f_3 N_3 + h(t) \tag{3.38}$$

The solution of the coupled equations (3.35), (3.36) and (3.38) is

$$N_1 = \left(\frac{\alpha f_3 N_{3i}}{(f_1 - f_3)}\right) e^{-f_3 t} + \alpha f_3 N_{3i} e^{-f_1 t}\left(h(t)\frac{e^{f_1 t}}{f_1} - \int h'(t)\frac{e^{f_1 t}}{f_1} dt\right) + c e^{-f_1 t} \tag{3.39}$$

where c is a constant and all other parameters are as previously defined.

If the signal from the deep trap h(t) is transient, it can be approximated to a constant, h, hence

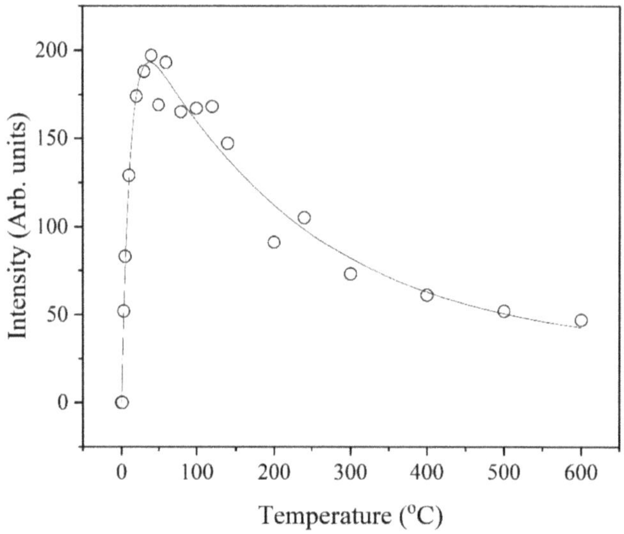

Figure 3.4. The time-response profile for PTTL of α-Al_2O_3:C for a system of one acceptor (peak 1) and two donors, the second of which only makes a negligible contribution to the PTTL. The line is a fit of equation (3.41) [18].

$$N_1 = \left(\frac{\alpha f_3 N_{3i}}{(f_1 - f_3)}\right)(e^{-f_3 t} - e^{f_1 t}) + \alpha f_3 N_{3i} \quad \frac{h}{s_1}(1 - e^{-f_1 t}) \tag{3.40}$$

which has the general form

$$N_1 = B(e^{-f_3 t} - e^{-f_1 t}) + D(1 - e^{-f_1 t}) \tag{3.41}$$

where $B = \alpha f_3 N_{3i}/(f_1 - f_3)$ and $D = \alpha f_3 N_{3i} h/f_1$. As an illustration, figure 3.4 shows a fit of equation (3.41) to the time-response profile for a system of one acceptor and two donors, the second of which only makes a minor contribution to PTTL.

3.8 Influence of stimulation temperature on PTTL intensity

The intensity of PTTL depends on various factors including dose and duration of illumination used to induce the phototransfer. The PTTL intensity should also be affected by measurement temperature since the amount of luminescence that can be optically stimulated depends on the stimulation temperature [8, 20]. In such cases, the luminescence intensity often goes through a maximum with measurement temperature. The initial increase is ascribed to thermal assistance to optical stimulation and the subsequent decrease, to thermal quenching. On the other hand, PTTL measurements have usually been carried out with samples being illuminated at room temperature or at low temperature. If the influence of thermal assistance is important, the intensity of luminescence phototransferred at higher temperatures should exceed that measured at room temperature. This is a direct result of phonon contribution to the photon-mediated release of trapped electrons by light.

3.8.1 Thermal assistance and thermal quenching

We assume a direct link between changes in the intensity of optically stimulated luminescence and PTTL intensity since the OSL appears during the process of phototransfer. If the emission of luminescence is affected by thermal assistance, the total probability of emission $1/\tau$ is influenced by a Boltzmann term $\exp(-E_a/kT)$ where E_a is the activation energy of thermal assistance [20]. The thermal assistance at one electron trap is independent of that at any other electron trap. Therefore, the overall probability of optical stimulation for n electron traps is

$$\frac{1}{\tau} = \left[\frac{1}{\tau_{\text{rad}}} + \nu \exp\left(-E_q/kT\right)\right] \prod_i^n \exp\left(-E_{ai}/kT\right) \tag{3.42}$$

where τ_{rad} is the radiative lifetime at absolute zero of temperature, ν and E_q are the frequency factor and activation energy for thermal quenching, and E_{ai} is the activation energy for thermal assistance at the ith electron trap [20]. The temperature dependence of the luminescence from all n electron traps is therefore

$$I(T) = \frac{I_o \prod_i^n \exp(-E_{ai}/kT)}{1 + C \exp\left(-E_q/kT\right)} \tag{3.43}$$

where $C = \nu\tau_{\text{rad}}$ [21]. Assuming a process involving a single electron trap or a single dominant electron trap in the case of closely spaced ones, the dependence of luminescence intensity (and so PTTL intensity) on temperature can be written as

$$I(T) = \frac{I_o \exp\left(-E_a/kT\right)}{1 + C \exp\left(-E_q/kT\right)} \tag{3.44}$$

Thermal assistance and thermal quenching can be simultaneously quantified using equation (3.44). Thermal quenching, which refers to incidences of non-radiative transitions, was discussed in section 1.6. Although the dependence of OSL or PTTL intensity can be used to demonstrate thermal quenching and thermal assistance, showing the effect of the latter alone is less easy to achieve. This is because the effect of thermal assistance to optical stimulation is often obscured by high intensity transient emissions. A way to demonstrate the effect is to compare the evolution of the intensity measured under optical stimulation alone with that obtained under simultaneous optical stimulation and heating as exemplified previously for quartz [22] or $Sr_4Al_{14}O_{25}$: Eu_{2+}, Dy_{3+} [23]. An example of such a comparison is shown in figure 3.5.

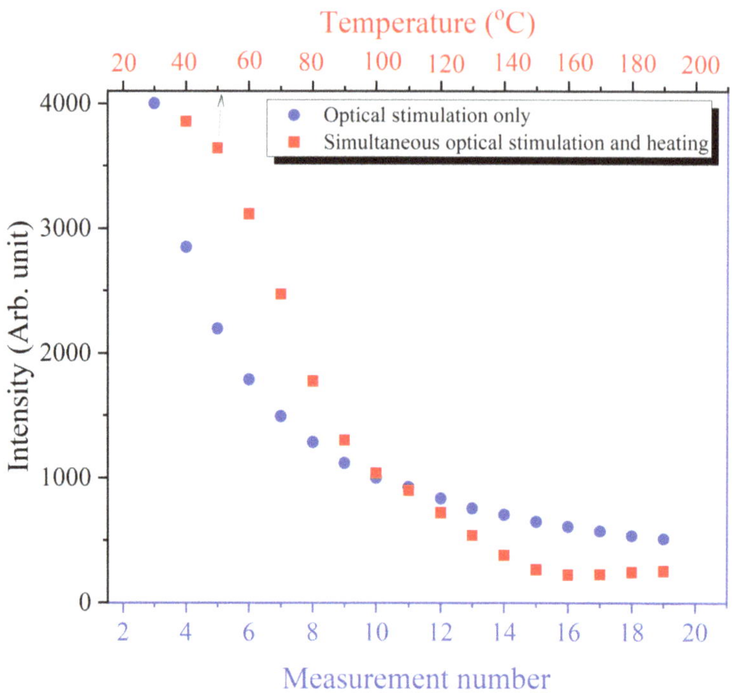

Figure 3.5. The change of luminescence intensity optically stimulated from $Sr_4Al_{14}O_{25}$: Eu^{2+}, Dy^{3+} using 870 nm infrared light at room temperature compared with that measured with the temperature increasing at 10°C intervals up to 180°C. Reprinted from [23].

3.9 Definition of PTTL

The mainstream understanding of phototransfer as it relates to TL is that if the TL measured off one electron trap is a result of electrons optically transferred from another more stable one, that thermoluminescence is 'phototransferred'. Thus, the photransfer acts in lieu of dose since the process is reliant on photons releasing electrons from donors. Phototransfer itself is influenced by the temperature at illumination used to induce it. The glow curve measured as a means to monitor the PTTL is just a temperature-resolved set of peaks. We therefore define PTTL as *a phonon-mediated temperature-and-wavelength resolved emission that arises at one electron trap due to a combined phonon-aided and photon-mediated release of electrons from another electron trap*.

3.10 Summary

Since PTTL is a TL process, it can be described using kinetic formulations. The relevant rate equations are non-linear and one invokes a battery of assumptions to obtain analytical solutions which can be applied to experimental data. An alternative method is to describe the PTTL only at the illumination stage using a phenomenological model in which assumptions are tested by experiment. To get a sense of the general expected behaviour, vector fields are instructive. Yet another approach is to obtain and display unique numerical solutions of rate equations. Such simulation has been used extensively for TL but less so for PTTL. This chapter has focused only on the influence of duration of illumination on PTTL intensity. As the PTTL intensity is affected by the illumination temperature, parts of the chapter are devoted to the analysis of the temperature-induced effects on phototransfer.

References

[1] McKeever S W S 1985 *Thermoluminescence of Solids* (Cambridge: Cambridge University Press)

[2] Chen R and McKeever S W S 1997 *Theory of Thermoluminescence and Related Phenomena* (Singapore: World Scientific)

[3] McKeever S W S, Moscovitch M and Townsend P D 1995 *Thermoluminescence Dosimetry Materials: Properties and Uses* (Nuclear Technology Publishing)

[4] McKeever S W S 2022 *A Course in Luminescence Measurements and Analyses for Radiation Dosimetry* (New York: Wiley)

[5] Pagonis V, Kitis G and Furetta C 2006 *Numerical and Practical Exercises in Thermoluminescence* (Berlin: Springer)

[6] Pagonis V 2022 *Luminescence Signal Analysis Using Python* (Berlin: Springer)

[7] Stoddard A E 1960 Effects of illumination upon sodium chloride thermoluminescence *Phys. Rev.* **120** 114

[8] Botter-Jensen L, McKeever S W S and Wintle A G 2003 *Optically Stimulated Luminescence Dosimetry* (Amsterdam: Elsevier)

[9] Alexander C S and McKeever S W S 1998 Phototransferred thermoluminescence *Phys J Appl. Phys.* **31** 2908

[10] Wintle A G and Murray A S 1997 The relationship between quartz thermoluminescence, photo-transferred thermoluminescence, and optically stimulated luminescence *Radiat. Meas.* **27** 611–24

[11] Chithambo M L, Niyonzima P and Kalita J M 2018 Phototransferred thermoluminescence of synthetic quartz: Analysis of illumination-time response curves. *J. Lumin.* **198** 146–54

[12] Furetta C 2003 *Handbook of Thermoluminescence* (Singapore: World Scientific)

[13] Moscovitch M 2011 The principles of phototransferred thermoluminescence *AIP Conf. Proc.* **1345** 323–34

[14] Edwards C H and David E P 2019 *Elementary Differential Equations* (Pearson Modern Classics)

[15] Chithambo M L 2021 Phototransferred thermoluminescence of tanzanite: a matrix-based analysis of time-response profiles and competition effects *J. Lumin.* **234** 117969

[16] Chithambo M L and Kalita J M 2021 Phototransferred thermoluminescence of BeO: time-response profiles and mechanisms *J. Appl. Phys.* **130** 195101

[17] Chen R and Pagonis V 2011 *Thermally and Optically Stimulated Luminescence: A Simulation Approach* (New York: Wiley)

[18] Chithambo M L, Seneza C and Kalita J M 2017 Phototransferred thermoluminescence of α-Al_2O_3:C: experimental results and empirical models *Radiat. Meas.* **105** 7–16

[19] Chithambo M L 2022 Phototransferred thermoluminescence of CaF_2: principles, analytical methods and mechanisms *J. Appl. Phys.* **132** 055102

[20] Chithambo M L 2018 *An Introduction to Time-resolved Optically Stimulated Luminescence* (Morgan & Claypool Publishers)
[21] Chithambo M L 2007 The analysis of time-resolved optically stimulated luminescence. II: Computer simulations and experimental results *J. Phys. D: Appl. Phys.* **40** 1880–89
[22] Chithambo M L and Galloway R B 2001 On the slow component of luminescence stimulated from quartz by pulsed blue light emitting diodes *Nucl. Instrum. Meth. B.* **183** 358–68
[23] Chithambo M L 2021 Thermal assistance in the optically stimulated luminescence of superluminous $Sr_4Al_{14}O_{25}$: Eu^{2+}, Dy^{3+} *Physica B: Condens. Matter.* **603** 412722

Chapter 4

Synthetic materials

Except for a few examples, most synthetic dosimeters luminesce brightly under optical or thermal stimulation. This chapter presents the phototransferred thermoluminescence (PTTL) of selected synthetic materials. The descriptions cover the measurement, analysis and mechanisms of the PTTL. The cases discussed are not meant to be exhaustive and are for the most part chosen owing to ease of access to data.

4.1 Synthetic quartz

The PTTL of synthetic quartz has seldom been studied systematically and there are only a few reports to draw from (e.g. [1–5]). This opening part looks at some conclusions reached from qualitative analysis of its PTTL.

4.1.1 Introduction

Phototransferred TL in quartz has historically been linked to selected electron traps, notably those at '325 °C' [6, 7], or at '375 °C' [8] as well as deep electron traps that are activated beyond 500 °C [9, 10]. In these examples, the PTTL is deduced to occur due to a transfer of electrons from a known or assumed donor to an acceptor electron trap. Despite the specificity about the donor adopted in this way, quartz has a range of optically stimulable electron traps [11], any of which may act as a donor for the PTTL. The description of the phototransfer phenomenon in terms of a select few electron traps should therefore be a cause for reassessment. As a case in point, optically stimulated luminescence (OSL), which has a direct bearing on the PTTL, can be obtained in various ways such as immediately after irradiation, after drawn out illumination (e.g. [12, 13]), at a range of stimulation temperatures, and even after a glow curve is depleted of most or all of its glow peaks (e.g. [5, 14]). It is therefore more instructive to analyse PTTL holistically in terms of a system of acceptors and donors as informed by experiment rather than to assume *ab initio* what the donors are. The number of donors can change with preheating. Indeed, preheating also affects their importance as do other experimental factors such as illumination.

doi:10.1088/978-0-7503-3831-8ch4

We start with synthetic quartz in this discussion as this material provides a good case on which general and systematic methods have been explored.

4.1.2 Conventional thermoluminescence

A glow curve of crystalline synthetic quartz [4] is shown in figure 4.1. There are three discernible peaks; an intense one near 130 °C collocated with a poorly

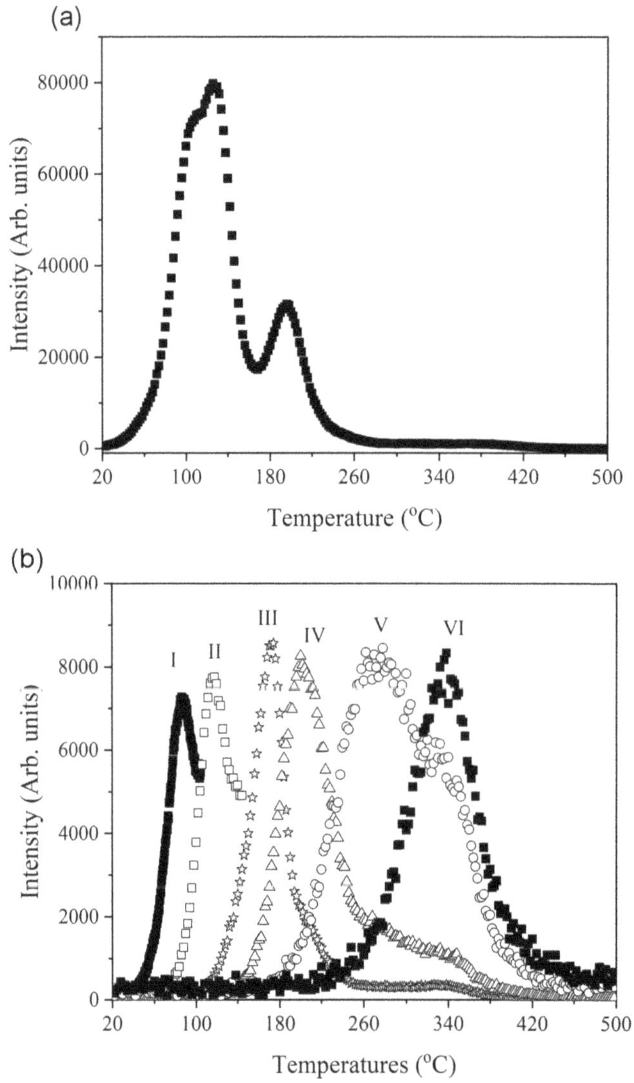

Figure 4.1. A glow curve of synthetic quartz recorded at 5 °C s^{-1}. Components of the glow curve abstracted using 'thermal cleaning' (b). The original data for the latter have been scaled up to better identify each component peak. The thermoluminescence was detected between 250 nm and 390 nm. The quartz, which was irradiated to 100 Gy in these measurements, was supplied by Sawyer Research Products, Ohio, USA. Reprinted from [4], Copyright (2018), with permission from Elsevier.

defined component at its lower temperature end and a third peak beyond 180 °C. Thermal cleaning reveals up to six peaks distributed throughout the glow curve (figure 4.1(b)).

4.1.3 Properties of PTTL

The PTTL of the quartz in question here, as induced by 470 nm blue light, displays contrasting behaviour depending on several factors. We single out the effect of preheating, look at the influence of duration of annealing on the PTTL and touch on the role of deep electron traps as donors in the process.

4.1.3.1 Influence of preheating on PTTL

Examples of glow curves obtained after preheating to selected temperatures are displayed in figure 4.2. When peak I near 130 °C is the only one removed by the preheating (curve *y*), illumination does not induce any PTTL. Phototransfer only ensues if the first three peaks have been removed (curve *z*). This result is telling because it suggests that the electron trap for peak III impedes emission of PTTL at peaks I and II by acting as a competitor for electrons released from donors. PTTL is also observed after preheating that removes peaks IV and V, or all six peaks [4]. The source of the PTTL once all peaks have been cleared must be deep electron traps.

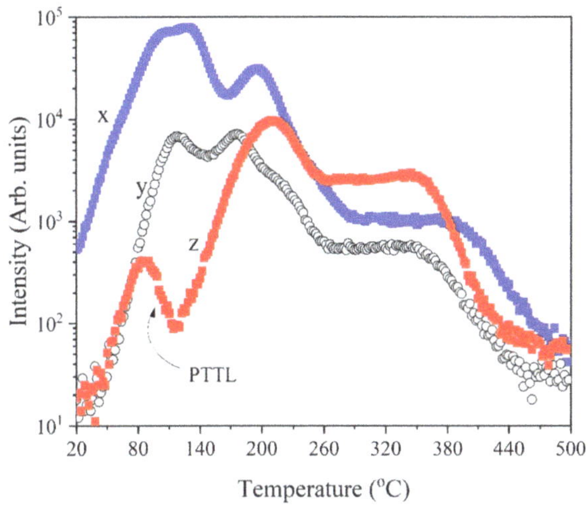

Figure 4.2. Glow curves obtained during preparatory tests for PTTL. A glow curve measured soon after irradiation (curve *x*) can be compared with one recorded after preheating to remove peaks I–III and illumination (curve *z*). Illumination induces PTTL only after a certain number of peaks have been removed. This is exemplified by the glow curve corresponding to preheating to 100 °C intended to remove peak I only (curve *y*). Reprinted from [4], Copyright (2018), with permission from Elsevier.

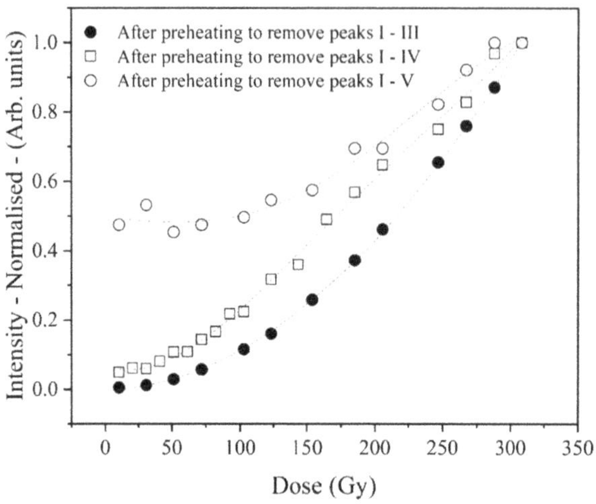

Figure 4.3. The dependence of intensity on dose for the PTTL peak observed after preheating to remove peaks III, IV and V, respectively (i.e. following preheating to 185, 250 and 300 °C). The lines through the data are visual guides. Reprinted from [4], Copyright (2018), with permission from Elsevier.

4.1.3.2 Influence of irradiation dose on PTTL intensity

The influence of irradiation dose on intensity of the PTTL is informative as it is an element of retrospective dosimetry. For illustration, we consider features of the peak induced under PTTL after preheating to remove each of peaks III, IV and V in that order. The growth curves for the PTTL for each of these three cases is shown in figure 4.3.

The growth curves in figure 4.3 relate to the relative concentration of electrons and holes at charge traps. The acceptor trap is depleted when the irradiated sample is preheated but then partially filled during the subsequent illumination. This process upsets the balance of charge in that the concentration of holes at the recombination centre always exceeds that of electrons phototransferred to the acceptor trap. When the irradiation dose is increased, the concentration of electrons phototransferred to the shallow trap also goes up. This causes the PTTL to correspondingly increase.

4.1.3.3 Dependence of PTTL intensity on the duration of illumination

The dependence of PTTL intensity on the duration of illumination for synthetic quartz follows many patterns. Figure 4.4(a) shows measurements of Bertucci *et al* [9] in (a) and of Chithambo and Niyonzima [4] in (b). Figure 4.4(a) is for analytical grade quartz preheated to 700 °C and 800 °C and is included here for its historical value. The very fact that emission was observed after preheating to such high temperatures is what prompted Bertucci *et al* [9] to attribute the PTTL to deep electron traps. Figure 4.4(b), which corresponds to preheating to 450 °C, can be interpreted in similar terms. The role of deep electron traps in defining PTTL in synthetic quartz should therefore be carefully considered.

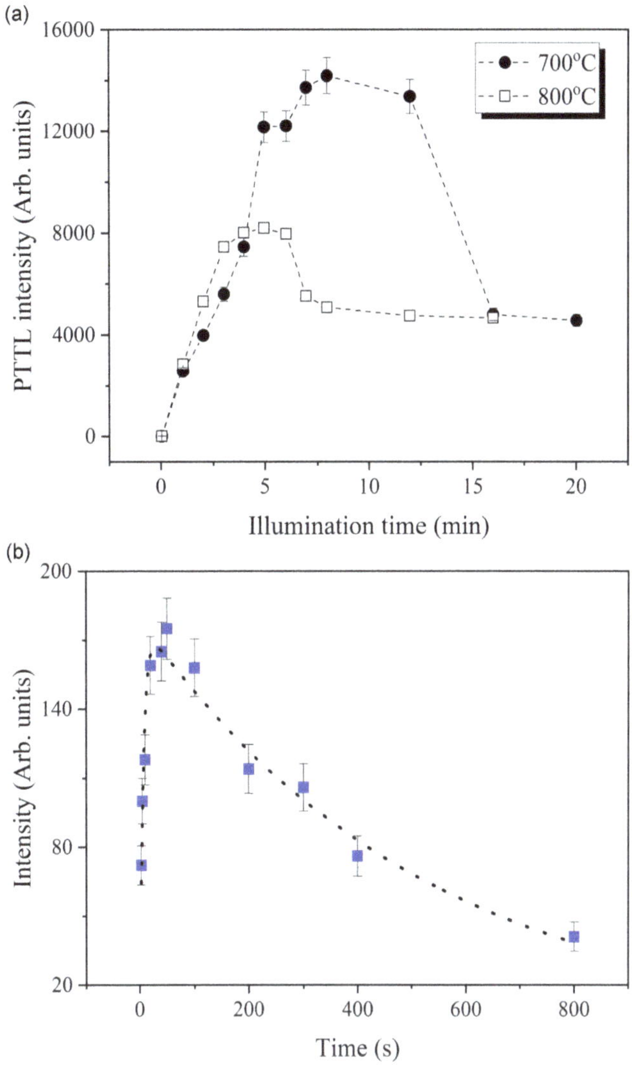

Figure 4.4. PTTL measurements as reported by Bertucci *et al* [9] (a) (reprinted from [9], Copyright (2011), with permission from Elsevier) and Chithambo and Niyonzima [4] (b) (reprinted from [4], Copyright (2018), with permission from Elsevier). Both refer to electron transfer from deep electron traps.

4.2 Annealed synthetic quartz

It is useful to consider the PTTL of the same synthetic quartz (Sawyer Research Products, Ohio, USA) when annealed. Various effects occur when quartz is heated to elevated temperatures. The heating inverts its phase [15], alters its thermoluminescence and optically stimulated luminescence sensitivity [11, 16], changes its luminescence lifetimes [16–18] and modifies the radioluminescence emission bands [19, 20]. Indeed, annealing accentuates radioluminescence emission bands with some becoming more prominent than others depending on the annealing temperature [19]. Our concern here is the nature of PTTL in annealed synthetic quartz, and how to interpret and analyse its time-response profiles. We refer to synthetic quartz annealed at 900 °C [5].

4.2.1 Conventional thermoluminescence

Glow curves of conventional thermoluminescence of synthetic quartz annealed at 900 °C for 10, 30 and 60 min are shown in figure 4.5. There are three stand-out peaks near 90, 120 and 180 °C. Thermal cleaning reveals three more weaker-intensity glow peaks [5]. The involvement of this collection of peaks in the PTTL is an important part of this discussion.

4.2.2 Properties of PTTL

Table 4.1 lists peak positions, preheating temperatures and identity of PTTL peaks for reference. The samples annealed for 10 min, 30 min and 60 min are referred to as samples A, B and C, respectively.

When sample A is preheated to remove only the first peak (I), or the first two peaks (I–II) or indeed the first three (I–III), no PTTL appears. It only appears when at least the first four peaks (I–IV) have been glowed out. Examples of PTTL peaks that come up after preheating to remove peaks I–V and I–VI are shown in figure 4.6. These are labelled P1 and P3.

When measurements are made on sample B instead, only peak P1 is obtained under phototransfer after removal of peaks I–IV and I–V by preheating. On the other hand, peaks P1, P2 and P3 are observed in sample C when peaks I–V are preheated off the glow curve. The tally of peaks that reappear under phototransfer depends on the annealing temperature. Three do so in sample C, two in sample

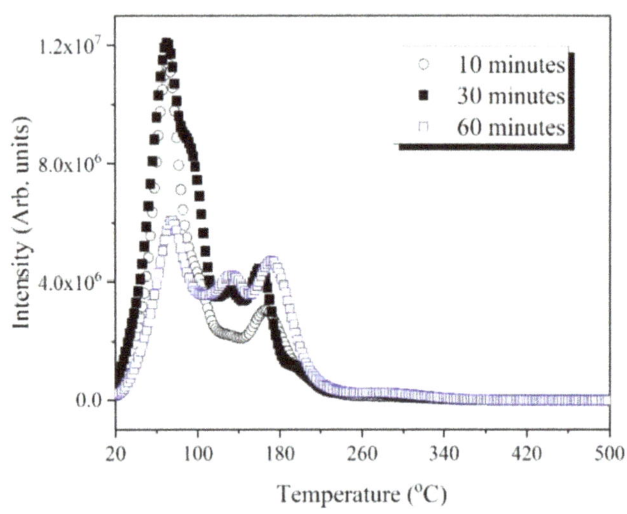

Figure 4.5. Glow curves measured at 1 °C s^{-1} from synthetic quartz annealed at 900 °C for 10, 30 and 60 min. A high dose of 200 Gy was used to facilitate the measurement of PTTL. Reprinted from [5], Copyright (2020), with permission from Elsevier.

Table 4.1. Thermal cleaning identified six peaks which are listed here together with their positions and preheating temperatures used in the PTTL measurements. PTTL peaks corresponding to samples A, B and C are denoted P1, P2 and P3.

	Thermoluminescence				Phototransferred thermoluminescence		
Sample	A	B	C		A	B	C
Peak	T_m (°C)	T_m (°C)	T_m (°C)	Preheat (°C)	Peak		
I	90	80	80	120			
II	122	110	120				
III	176	136	134	180			
IV	210	196	188	250	P1 P3	P1	
V	240	240	235	350	P1 P3	P1	P1 P2 P3
VI	340	330	340	500	P1 P3		

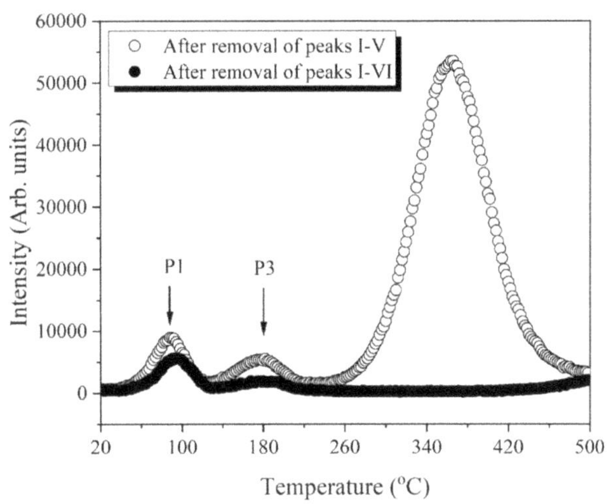

Figure 4.6. Glow curves measured off sample A after preheating to remove peaks I–V and I–VI. The preheating temperatures used are listed in table 4.1. The PTTL peaks induced by 470 nm light are denoted P1 and P3 and correspond to TL peaks I and III obtained under conventional TL. Reprinted from [5], Copyright (2020), with permission from Elsevier.

A and only one in sample B. We also alluded to the importance of deep electron traps in defining the PTTL in synthetic quartz. Further examination of this aspect [5] suggests that whether PTTL linked to deep electron traps is obtained depends on the annealing temperature. Such PTTL is observed in sample A but not when the quartz is annealed for longer than the 10 min of sample A.

4.2.3 Pulse annealing

The role of electron traps as donors or acceptors during PTTL can be investigated by systematically removing each peak in the glow curve portion by portion and

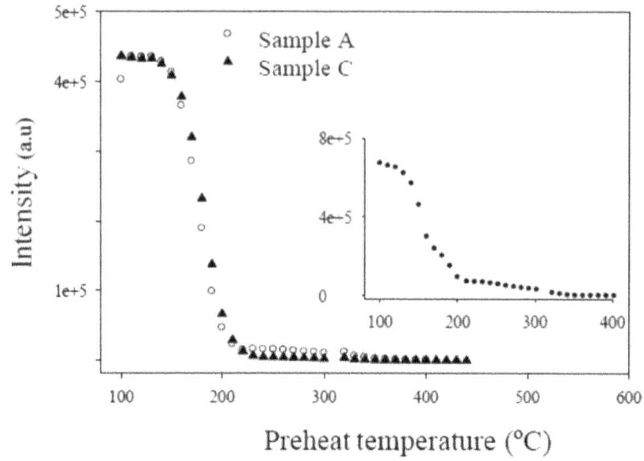

Figure 4.7. The intensity of peak P1 against preheating temperature for samples A and C and, in the inset for sample B. Reprinted from [5], Copyright (2020), with permission from Elsevier.

monitoring the intensity of any PTTL peaks as well as that of any peaks not affected by the preheating. This partial heating method is akin to 'pulse annealing' [21]. An irradiated sample is first preheated to deplete specific electron traps. The sample is then illuminated to phototransfer some electrons from deeper traps to emptied shallower ones. The process is normally assumed to occur via the conduction band. PTTL is monitored when the material is reheated. For brevity, we discuss the application of the method on peak P1 only.

Figure 4.7 shows the influence of preheating on the intensity of peak I for samples A and C, and in the inset for sample B [5]. If the preheating temperature lies below the position of a particular peak, all glow peaks measured after illumination relate to conventional TL, hence the reference to peak I to start with. If, on the other hand, the preheating goes beyond the peak position and yet the peak still comes up, then what is measured is PTTL. Thus, a plot like figure 4.7 shows how the intensity of the peaks respond to preheating.

The luminescence intensity is not affected by preheating between, say, 100 °C and 120 °C because the preheating does not deplete any putative donor. Such depletion can only occur if the preheating is extended further as is evident in the sharp decrease in intensity. The donors in question are presumably the electron traps for peaks II and III. The role of deep electron traps is clearly minimal.

4.2.4 Influence of heating rate on PTTL intensity

Peaks P1 and P3 for samples A and C decrease in intensity with heating rate in a manner consistent with thermal quenching [5]. If, in measurements carried out at different heating rates, the intensity I_u (in units of counts/K) obtained at the slowest heating rate is assumed to be least affected by thermal quenching, then its values are related to intensities $I_q(T)$ recorded at faster heating rates as

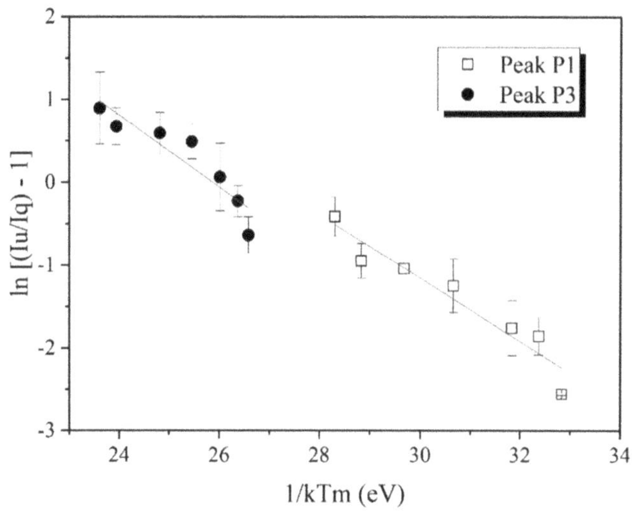

Figure 4.8. Plots of ln $[(I_u/I_q) - 1]$ against $1/kT_m$ used to study thermal quenching using peaks P1 and P3. The measurements correspond to preheating which removes the first four of the six peaks in the glow curve. Reprinted from [5], Copyright (2020), with permission from Elsevier.

$$I_q(T) = \frac{I_u}{1 + C\exp(-\Delta E/kT_m)}, \tag{4.1}$$

where T_m is the peak maximum, $C = \nu\tau_{\text{rad}}$, where ν is the frequency factor for the non-radiative process, τ_{rad} is the radiative luminescence lifetime [18], ΔE is the activation energy for thermal quenching and k is Boltzmann's constant. Thermal quenching can be quantified by plotting ln $[(I_u/I_q) - 1]$ against $1/kT_m$ [22] as exemplified in figure 4.8. The resulting values of the activation energy of thermal quenching are 0.62 ± 0.04 eV using peak P1 and 0.65 ± 0.05 eV using peak P3 in sample A. Comparative analysis using samples B and C gives 0.63 ± 0.02 eV and 0.66 ± 0.03 eV. These values are all typical of synthetic quartz (e.g. [23]) and show that the emission of luminescence occurs at a common recombination centre. This is consistent with radioluminescence spectra from the same quartz [19], which show that varying the duration of annealing only affects the intensity not the position of emission bands.

4.3 α-Al_2O_3:C

4.3.1 Introduction

α-Al_2O_3 is a phase of aluminium oxide that crystallizes into a rhombohedral lattice with trigonal symmetry. α-Al_2O_3 has unique physical and chemical properties which generate point defects that can be exploited in luminescence-based applications [24–26]. In particular, colour centres can be created in the oxygen sub-lattice in various ways including by incorporation of impurities, bombardment by energetic neutrons, ion or electron beams [24–27]. Oxygen vacancies in α-Al_2O_3 enhance the formation of colour centres, the principle of which is the F centre, an oxygen vacancy with two trapped electrons or alternatively one with a single trapped electron, the F^+ centre [24–28].

When α-Al_2O_3 is doped by carbon to produce α-Al_2O_3:C, a luminescence dosemeter, α-Al_2O_3 is chemically reduced. This causes a large increase in the concentration of F and F^+ centres. These colour or electron centres and possibly their aggregates serve as luminescence sites for electrons optically stimulated from electron traps in α-Al_2O_3:C. A particularly intense glow peak and several weaker intensity others in the glow curve of α-Al_2O_3:C reflect the diverse role of electron traps in α-Al_2O_3:C [27–38]. The presence of deep electron traps in α-Al_2O_3:C is implicit from the generation of PTTL in irradiated samples [35–37], by thermally-assisted optical stimulation [39] as well as by thermally-assisted time-resolved optical stimulation of luminescence [40].

The number of studies devoted to the mechanisms of thermoluminescence in α-Al_2O_3:C is extensive and yet comparatively few have explored PTTL for the same purpose. Colyott *et al* [36] used ultraviolet-blue light to induce PTTL in α-Al_2O_3:C and alluded to the point that emission of luminescence involves both electrons and holes. Indeed, in measurements made at 0.3 °C s^{-1}, they surmised that a hole trap activates at 277 °C and 630 °C. The PTTL of the main glow peak in α-Al_2O_3:C was investigated by Bulur and Göksu [37], who linked it with phototransfer from a deep electron trap near 500 °C. These authors asserted that no PTTL could appear if an irradiated sample was preheated to more than 600 °C.

A substantive discussion of phototransferred thermoluminescence in α-Al_2O_3:C was reported by Chithambo *et al* [35]. The study analysed the PTTL at all electron traps monitored in this material. The objectives included the effect on phototransfer of preheating temperature, the duration of preheating, examination of competition effects, the duration of illumination and, principally, the development of analytical methods for the time-response profiles. Chithambo *et al* [35] introduced the use of phenomenological mathematical models to analyse PTTL in terms of systems of an acceptor and donors where their number is decided, not by assumption, but by experimental results. The current discussion mostly draws from that report.

Models of PTTL for a conceptual system of one acceptor and one donor, i.e. two electron traps, were discussed by Alexander and McKeever [41]. We recall that the PTTL in natural quartz was thought of in the same terms by Wintle and Murray [6]. Alexander and McKeever [41] were concerned with explaining how the PTTL intensity evolves with duration of illumination. Changes caused by adding a second but non-radiative recombination centre were considered interesting enough for inclusion in the same report. Alexander and McKeever [41] used the numerical solutions to the systems of differential equations to explain various scenarios for relevant conceptual cases.

The glow curve of α-Al_2O_3:C corresponds to a succession of electron traps as exemplified by figure 1.2. Nevertheless, the models of Alexander and McKeever [41], developed for a system of two electron traps, one an acceptor and the other a donor, offer a thought guide on how to analyse PTTL in α-Al_2O_3:C. The literature on its TL invariably shows that the glow curve consists of an apparently single prominent peak and a number of less intense ones (e.g. [28–38]). All these candidates can influence phototransfer in one way or another. α-Al_2O_3:C is also deduced to have deep electron traps at 680–930 °C and deep hole traps at 530–600 °C [33, 36, 42]. Emission bands at 1.25, 1.49, 3.0, 3.27 and 3.8 eV, that is, near 990, 830, 410, 380 and 330 nm are characteristic of the host, Al_2O_3 [24, 26]. The ones that stand out in α-Al_2O_3:C are near 3.0 eV (410 nm) and 3.8 eV (330 nm), respectively [24, 26, 28, 43]. These bands are associated with transitions at F and F^+ electron centres. It is therefore not in question that α-Al_2O_3:C is a complex system as far as systematic analysis of its PTTL goes.

One of the recurring themes in the study of TL and PTTL is the extent to which one ought to rely on either generic or theoretical models or empirically-based ones. In their study, which was the first to systematically study PTTL, Alexander and Mckeever [41] cautioned against being too liberal with assumptions. This point is also relevant for experiment-based analysis. With judiciously chosen parameters for their test case, they [41] successively accounted for the archetypal case of a system with one donor. An obvious follow-up question that arises is how to address the issue of the PTTL of a system of multiple electron traps and hole centres such as α-Al_2O_3: C.

We now look at the measurement and analysis of phototransfered thermoluminescence of α-Al_2O_3:C. The role of electron traps as acceptors or donors changes with preheating. The influence of duration of illumination on the intensity of PTTL is first dealt with qualitatively and also addressed using an analytical phenomenological model intended to reflect experimental results. The PTTL corresponds to phototransfer using 470 nm blue light and refers to preheating between 100 °C and 800 °C [35].

4.3.2 Glow curve structure

Figure 4.9 (solid circles) is a glow curve measured at 5 °C s^{-1} after irradiation to 0.5 Gy. There are three glow peaks. The most intense emission (peak II) occurs between lower intensity responses (peaks I and III). This particular sample has two additional peaks, one embedded within the main peak at its higher temperature end whereas the other is a barely discernible component beyond peak III [30].

4.3.3 Properties of PTTL

The dependence of PTTL intensity on duration of illumination is shown in figure 4.10 for various preheating temperatures. Such profiles can be discussed as relating to a system of acceptor and donors where the number of the latter changes with preheating.

4.3.3.1 Phototransfer after removal of peak I only

The open symbols in figure 4.9 that we described earlier show a portion of a glow curve measured from an irradiated sample after preheating to 100 °C and illumination for 30 s [35]. This preheating removes only peak I which then reappears under phototransfer. The change in its intensity with duration of illumination is displayed in figure 4.10(a). The intensity goes through a maximum with

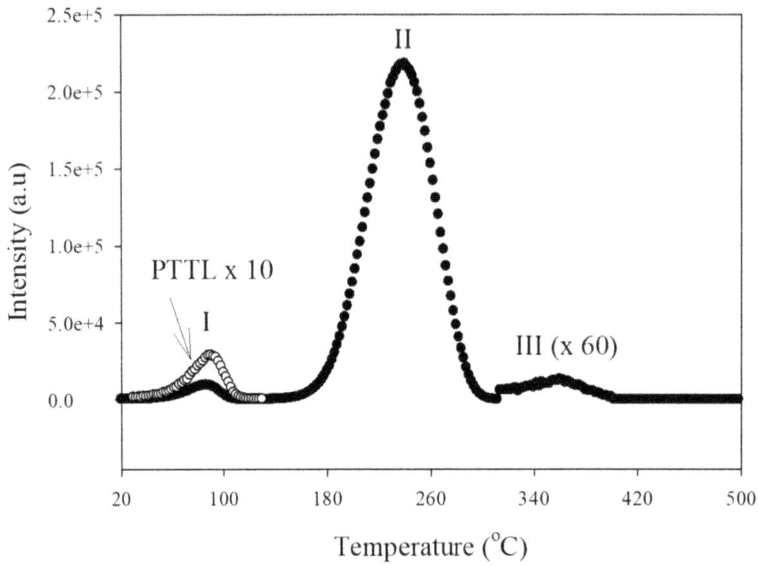

Figure 4.9. When α-Al_2O_3: C is heated at 5 °C s^{-1} after irradiation to 0.5 Gy, there are three nominal peaks in its glow curve as shown here. The extreme intensity of the main peak suppresses those of peaks I and III, which have had to be scaled up to 10 and 60 × respectively in this plot. The open symbols show peak I reproduced under phototransfer. Reprinted from [35], Copyright (2017), with permission from Elsevier.

illumination. This behaviour can be attributed to how charge transfer proceeds from deeper electron traps owing to optical stimulation. The PTTL becomes brighter with illumination because the number of electrons phototransferred to the electron trap of peak I exceeds any lost from the same trap during illumination. Further depletion of donors is then reflected as a continual drop in the intensity of the PTTL. This qualitative explanation applies if retrapping into the electron traps is negligible.

4.3.3.2 Phototransfer after removal of peaks I and II

When peaks I and II are removed by preheating, they are both reproduced by phototransfer. The time-response profiles are shown in figure 4.10(b) for peak I and, in the inset, for peak II.

4.3.3.3 Phototransfer after removal of the first three peaks

When preheating removes peaks I–III, only peaks I and II are reproduced under phototransfer. Figure 4.10(c) shows the dependence of their PTTL intensity on duration of illumination. A notable departure from the archetype for peak II (open symbols) is that the PTTL intensity beyond the maximum slowly approaches a stable value of higher intensity.

4.3.3.4 Phototransfer from deep electron traps

If all peaks are erased from the glow curve by preheating to 500 °C, illumination induces intense emission at peak II, none at peak III and hardly any at peak I.

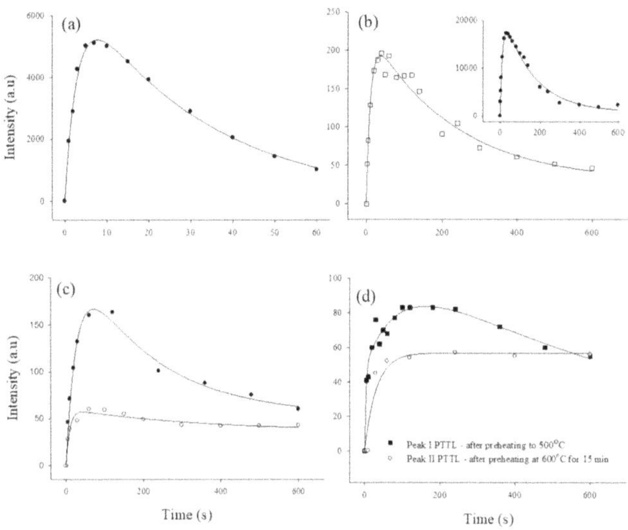

Figure 4.10. When the duration of illumination changes, so does the PTTL intensity. The time-response profiles are shown for peak I after preheating to 100 °C, which removes this peak only (a) for peak I and, in the inset, for peak II after preheating to 290 °C, which removes these two peaks before illumination (b) and for peak I (solid symbols) and peak II after removal of peaks I–III by preheating to 390 °C (c). The PTTL induced by phototransfer from deep electron traps is shown for peaks I and II corresponding to preheating to 500 °C and 600 °C, respectively. The more puzzling case of peak II after preheating to 500 °C is shown in figure 4.11. The lines through the data are mathematical descriptions of experimental behaviour as explained within the chapter. Reprinted from [35], Copyright (2017), with permission from Elsevier.

Figure 4.11 displays the plot for peak II. To get very bright PTTL after preheating to 500 °C is somewhat surprising. Chithambo *et al* [35] attributed this to release of holes from a hole trap when the sample is preheated this high after irradiation. The released holes recombine with pre-existing F centres thereby promoting the formation of F^+ centres. Since the recombination, at an F^+ centre, of some electrons phototransferred from the deep trap explains the PTTL, an increase in the concentration of F^+ centres is likely to lead to correspondingly intense PTTL. This is consistent with some observations, say, by Akselrod and Gorelova [42] or Colyott *et al* [36], whose results pointed to the existence of hole traps between 530 °C and 600 °C and at 280 °C (for heating at 0.3 °C s^{-1}), respectively. The dependence of PTTL intensity on duration of illumination after preheating to 500 °C for peaks I and II is shown in figure 4.10(d).

The measurements related to preheating to 600 °C, 700 °C and 800 °C also intended to sense deep electron traps are instructive. In this case, only peak II is reproduced under phototansfer but poorly so. The weak intensity is due to depletion of deep donor electron traps caused by preheating before any phototransfer. Figure 4.12 shows the PTTL recorded after preheating at 600 °C (solid symbols) and at 700 °C (open symbols) for 6 min.

Whereas preheating at 600 °C and 700 °C for 6 min does not change the typical PTTL intensity-time plot, lengthening the duration of preheating to

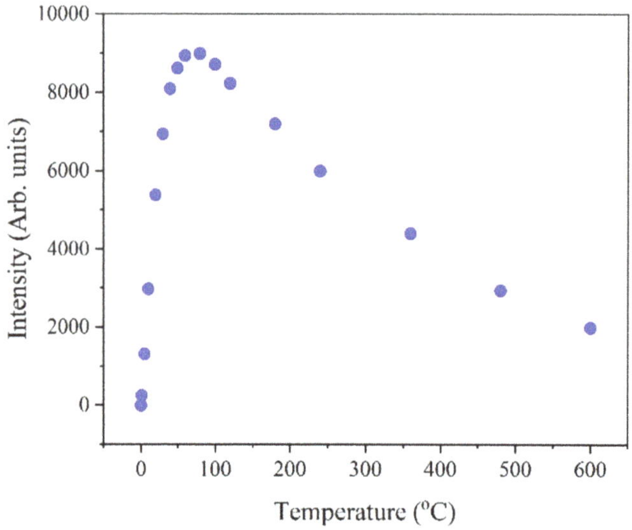

Figure 4.11. The PTTL monitored at peak II after preheating to 500 °C is counterintuitively intense. The change of the PTTL intensity with illumination as shown here is archetypal. Reprinted from [35], Copyright (2017), with permission from Elsevier.

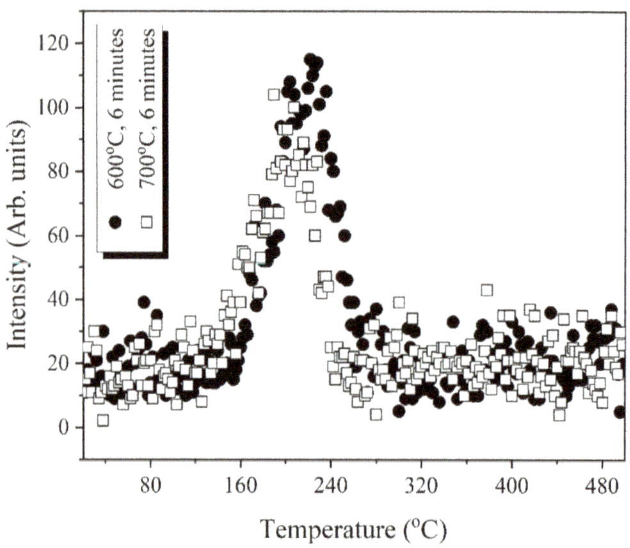

Figure 4.12. PTTL measured after preheating at 600 °C (solid symbols) and at 700 °C for 6 min (open symbols). Reprinted from [35], Copyright (2017), with permission from Elsevier.

15 min (or preheating at 800 °C for 6 min) causes the PTTL intensity to saturate with illumination time (figure 4.10(d)). It is likely that this is so because changing the illumination time has a negligible effect on the residual concentration of trapped electrons that can be phototransfered to induce PTTL

at the acceptor traps [35]. The important result here is that it is possible to still measure PTTL even where the preheating is well over 500 °C as in this example, at 800 °C.

4.3.4 Phenomenological model of PTTL

4.3.4.1 Qualitative basis

The PTTL of α-Al_2O_3:C can be accounted for with reference to the energy band scheme of figure 4.13. The shallow-, main- and intermediate-energy electron traps referring to peaks I, II and III are labelled ST, MT and IET, respectively. The representative deep hole trap and deep electron trap, denoted DHT and DT, are each activated only at elevated temperatures of 530–600 °C and 680–930 °C, respectively, using appropriate heating rates (e.g. [33, 36, 42]). The recombination centre is labelled R. When an irradiated sample is preheated to deplete shallow electron traps, illumination partially refills them by phototransfer from deeper electron traps. The optical removal of electrons from each donor or loss of electrons from the acceptor owing to illumination are indicated by the upward arrows. For this particular sample, retrapping is neglected because peaks I and III follow first order kinetics [30, 38] whereas for peak II, the order of kinetics is ~1.42 [35, 44].

4.3.4.2 Phenomenological model

The processes leading to production of PTTL in α-Al_2O_3:C can be modelled on the basis of systems of acceptor and donors. The number of donors depends on the

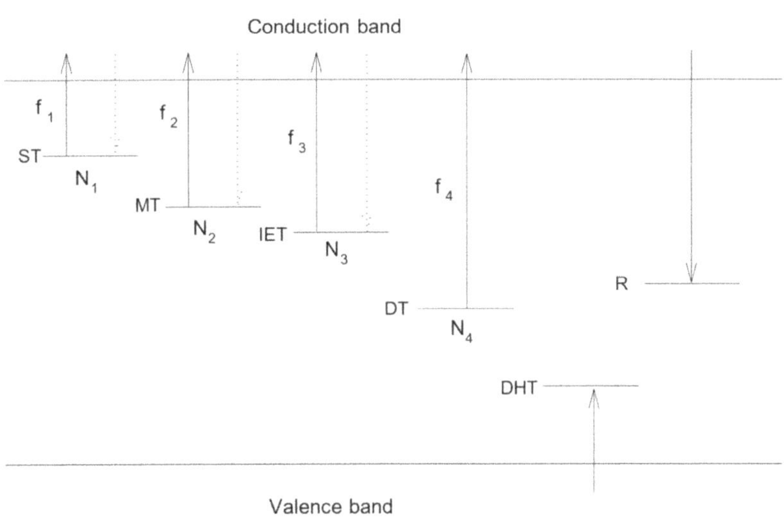

Figure 4.13. An energy band scheme used to describe PTTL in α-Al_2O_3:C. The shallow-, main- and intermediate-energy electron traps corresponding to peaks I, II and II are labelled ST, MT and IET. The deep electron trap (DT), deep hole trap (DHT) and the recombination centre (R) are included in the discussion. The downward dotted arrow represents the transfer, to a given acceptor, of a proportion of electrons optically released from a particular donor. Optical stimulation from each electron trap is shown by upward arrows. Reprinted from [35], Copyright (2017), with permission from Elsevier.

preheating temperature. For instance, if only peak I is removed by preheating, all electron traps activating at temperatures beyond it can serve as donors for any PTTL observed at peak I. If multiple peaks are removed, the number of donors scales down accordingly. The manner in which the PTTL intensity changes with illumination time can therefore be analysed using an analytical model where the rate equations are formulated in terms of the proportion of charge that ends up at the particular acceptor once optically released from each donor. The differential equations are formulated only at the illumination stage. The system for α-Al_2O_3: C under discussion here comprises four electron traps, one hole trap and one recombination centre. The rate equations are set up assuming that backscatter to acceptors is negligible and that some electrons are lost from all electron traps during illumination. The resulting systems of equation will have analytical solutions which can be applied to experimental data.

4.3.4.3 PTTL following removal of peak I only

When peak I is the only one removed by preheating, the donors for its PTTL can be any or all of electron traps for peaks II, III and the deep electron trap. Of these, the electron trap for the main peak is the most sensitive to optical stimulation. In comparison, phototransfer from either the electron trap of peak III or from the deep electron trap (DT) should be negligible. The PTTL can therefore pragmatically be considered in terms of a system of one acceptor (the electron trap for peak I) and one donor (the electron trap for peak II). However, if one chooses to fold in the role of all electron traps, the system of coupled rate equations describing electron transfer between donors and acceptor can be written as

$$\frac{dN_4}{dt} = -f_4 N_4 \tag{4.2}$$

$$\frac{dN_3}{dt} = -f_3 N_3 \tag{4.3}$$

$$\frac{dN_2}{dt} = -f_2 N_2 \tag{4.4}$$

$$\frac{dN_1}{dt} = -f_1 N_1 + a_2 f_2 N_2 + a_3 f_3 N_3 + a_4 f_4 N_4 \tag{4.5}$$

where f_i ($i = 1,\ldots,4$) are the probabilities of optical stimulation from the respective electron traps. These are level 2 for the main trap, level 3 for peak III and level 4 for the deep electron trap. The terms $a_i f_i N_i$ express the fact that only some of the electrons released from each source trap transit to the acceptor trap. The parameters a_i are constants of proportionality; N_1 is the concentration of electrons at the acceptor (ST) and the remaining terms N_i ($i = 2,\ldots,4$) are each the concentration of electrons at the donor electron traps at any time t. The time dependence of the occupancy at the acceptor electron trap is therefore

$$N_1 = A_1(e^{-f_2 t} - e^{-f_1 t}) + A_2(e^{-f_3 t} - e^{-f_1 t}) + A_3(e^{-f_4 t} - e^{-f_1 t}) \quad (4.6)$$

where the constants A_1, A_2 and A_3 are respectively proportional to N_{2i}, N_{3i} and N_{4i}, the initial concentration of electrons at the electron traps. If, on the other hand, the only contributions taken into account are those from the main electron trap, the system in question reduces to that of a single donor and one acceptor for which

$$N_1 = A_1(e^{-f_2 t} - e^{-f_1 t}) \quad (4.7)$$

We assume as before [6] that the PTTL intensity will be proportionate to the concentration of electrons at the acceptor trap at the end of the illumination.

4.3.4.4 PTTL from deep electron traps

When the sample is illuminated after preheating to 500 °C, only deep electron traps should be involved in phototransfer as donors. The PTTL associated with phototransfer from deep traps is, in many materials, typically weak. However, results for α-Al_2O_3:C offer interesting contrasts. There is no PTTL at peak III, barely any from peak I and yet that from peak II is particularly intense [35]. The time-response profile of peak I corresponding to preheating to 500 °C is shown in figure 4.10(d). The phototransfer relates to transfer from a single donor, namely, the deep electron trap. Thus, one would expect the profile to resemble that of a system of one acceptor and one donor as exemplified in figure 4.10(a). However, the experimental result (figure 4.10(d)) appears otherwise. The growth and decrease of the PTTL intensity with illumination is considerably slow. This behaviour suggests that there is possibly another component to the expected phototransfer from the deep electron trap. The origin of this additional contribution is not obvious. However, it is known that electron trap for peak III also acts as a donor [35]. Thus, it is likely that some electrons stimulated from this electron trap end up at the electron trap for peak I. Although such transfer is negligible, it is nevertheless important because any contribution from a deep electron trap does not supplement it to any significant extent.

The transport of electrons leading to PTTL at peak I can therefore be written as

$$\frac{dN_4}{dt} = -f_4 N_4 \quad (4.8)$$

$$\frac{dN_3}{dt} = -f_3 N_3 + \delta f_4 N_4 \quad (4.9)$$

$$\frac{dN_1}{dt} = -f_1 N_1 + \gamma_3 f_3 N_3 + \gamma_4 f_4 N_4 \quad (4.10)$$

Equation (4.8) expresses the optical removal of electrons from the deep electron trap. Equation (4.9) describes the same movement for electron trap IET with the minor difference being that this electron is empty at the start of illumination. Equation (4.10), where δ, γ_3 and γ_4 are constants of proportionality, expresses the point that the electron transition at the acceptor includes inward transfer of some electrons from donors DT and IET. The solution of equations (4.8)–(4.10) in terms of N_1 is

$$N_1 = A^*(e^{-f_4 t} - e^{-f_1 t}) - B^*(e^{-f_3 t} - e^{-f_1 t}) + C^*(e^{-f_4 t} - e^{-f_1 t}) \quad (4.11)$$

where $A^* = \gamma_3 \delta f_3 f_4 N_{4_i} / [(f_1 - f_4)(f_3 - f_4)]$, $B^* = \gamma_3 \delta f_3 f_4 N_{4_i} / [(f_1 - f_3)(f_3 - f_4)]$ and $C^* = \gamma_4 f_4 N_{4_i} / (f_1 - f_4)$. The line through the data in figure 4.10(d) (solid symbols) is a fit of equation (4.11).

4.3.4.5 PTTL from very deep electron traps

When an irradiated sample is preheated at 600 °C, 700 °C or 800 °C, the deep electron trap becomes so depleted that the number of electrons phototransferred to the acceptor electron trap MT, the only one for which PTTL appears, no longer scales with duration of illumination. The optically-induced removal of electrons from the deep trap can be expressed as $N_4 = N_{4i} e^{-f_4 t}$. Since preheating leaves the occupancy at the deep trap negligible and because $f_4 \ll 1$, it follows that $N_4 \approx N_{4i}(1 - f_4 t) \approx N_{4i}$ where all symbols are as defined before. The number of electrons optically transferred from the deep electron trap starts out as negligible and will remain as such despite illumination and can hence be approximated as a constant. The phototransfer to the electron of the main peak can therefore be written as

$$\frac{dN_2}{dt} = -f_2 N_2 + \gamma f_4 N_{4i} \quad (4.12)$$

from which

$$N_2 = \kappa(1 - e^{-f_2 t}) \quad (4.13)$$

where γ is a constant of proportionality and $\kappa = \gamma f_4 N_{4i}/f_2$. Equation (4.13) is a saturating exponential from the start of illumination at $t = 0$. Equation (4.13) is applied in figure 4.10(d) to the plot for PTTL measured at peak II after preheating at 600 °C for 15 min.

4.3.5 Effect of thermal quenching on PTTL peak I

Whether measured under phototransfer [35] or by conventional means (e.g. [30, 38]), peak I is known to follow first order kinetics, to have an activation energy ~0.7 eV and a frequency factor of the order of 10^{10} s^{-1}. The electron trap for peak I is thought to be the C_{Al} site [45] and has noteworthy features related to thermal quenching.

The intensity of PTTL peak I decreases with heating rate as might be expected when thermal quenching is important (figure 4.14). This is not observed in comparative measurements under conventional TL [30]. To explain the difference, we invoke the principle that the luminescence process is a combination of radiative and non-radiative events, that is,

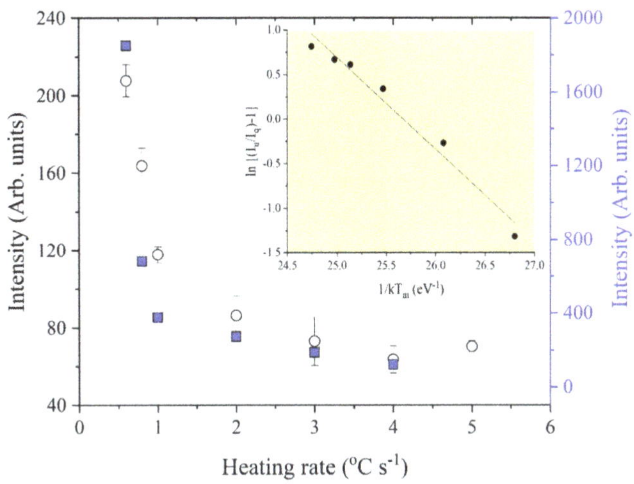

Figure 4.14. The dependence on heating rate of the intensity of peak I obtained under phototransfer after irradiation to 0.5 Gy (open circles) and of the same peak measured as conventional TL after irradiation to 0.1 Gy (solid squares). A plot of $\ln[(I_u/I_q) - 1)]$ against $1/kT_m$ used to quantify thermal quenching (inset). The intensity in the main graph is plotted in unconverted units for illustration. Analytical methods for such data is otherwise discussed in Ref. [23]. Reprinted from [35], Copyright (2017), with permission from Elsevier.

$$\frac{1}{\tau} = \frac{1}{\tau_{\text{rad}}} + \frac{1}{\tau_{\text{th}}} \exp\left(-\frac{\Delta E}{kT}\right), \tag{4.14}$$

where $1/\tau_{\text{rad}}$ is the probability of luminescent transitions, which is assumed to be independent of temperature and $1/\tau_{\text{th}}$ is the probability of thermal excitation where ΔE is the activation energy of thermal quenching [24, 46]. It is implicit from equation (4.14) that the luminescence intensity will depend on and should decrease with temperature. Studies on thermal quenching in α-Al_2O_3:C usually focus on its effect on the main peak (e.g. [27, 30, 38, 43, 46–49]).

Between the main peak and peak I, the latter is considerably weaker in intensity. If for peak 1 radiative transitions are far fewer than non-radiative ones, the intensity of the resulting luminescence can never offset the loss of signal due to the alternative non-radiative route and the TL intensity should tend to decrease with heating rate. This hypothesis, tested on Al_2O_3:C, showed that whereas the intensity decreases with heating rate when the TL is measured after irradiation to 0.1 Gy, it increases when a larger dose of 0.5 Gy is used [38], a behaviour also observed in quartz [23]. Analysis on the same principles as outlined in section 4.2.4 (figure 4.14, inset) gave $\Delta E = 1.03 \pm 0.08$ eV which compares favourably with say ΔE=1.025 ± 0.002eV [43], 1.045 ± 0.002eV [47] or 1.08 ± 0.03eV [48] obtained for α-Al_2O_3:C. This shows that electron traps associated with peaks I–III in α-Al_2O_3:C have a common recombination centre. Although the practice of using high-temperature glow peaks to quantify thermal quenching is entrenched, with judicious empiricism, the same can be achieved using peaks at temperatures near ambient.

4.4 BeO

4.4.1 Introduction

Beryllium oxide (BeO) is a chemically stable binary oxide. With a wide band gap of the order of 10.5 eV [50–52], BeO has long been of interest owing to its intrinsic and extrinsic point defects [24, 53], some of which act as luminescence sites. Although studies of the solid-state features of BeO are archival [54], it is the recording of intense ultraviolet luminescence from BeO [55] and development of methods for growing of its crystals [56, 57] that provided the impetus for the now voluminous literature on its luminescence. These studies tend to fall into two main themes, namely, those concerned with BeO as a prospective dosemeter [52, 55, 58–63], owing to its near-tissue equivalence, viz. effective atomic number = 7.13 [60], and efforts to explain the characteristics of its point defects. These two categories have seldom been reciprocal causing the literature on BeO to be replete with inconsistencies.

The crystal structure of BeO is hexagonal and resembles that of wurtzite [64]. A polyhedral model of the crystal structure is displayed in figure 4.15. Each beryllium atom is surrounded by a tetrahedron of oxygen atoms and, likewise, each oxygen atom by a tetrahedron of beryllium atoms. Although BeO attains a hexagonal close-packed structure, its c/a ratio at 1.6226 [53, 65] is somewhat less than the ideal 1.633, which is indicative of a slight distortion in its structure.

BeO contains colour centres such as F and F^+ types in its oxygen sub-lattice [24, 66]. In particular the F^+ centres can be accommodated at two dissimilar anion positions [67]. The positions of the F and F^+ bands have been calculated as at 4.7 eV and 3.7 eV, respectively, using density functional theory [68]. Besides these, other point defects such as hole centres have been identified by various means including electron spin resonance [24, 53].

Figure 4.15. A polyhedral model of the crystal structure of BeO is analogous to that of wurtzite shown here. Be^{2+} is bonded to four equivalent O^{2-}. In the same way, O^{2-} is linked to four equivalent Be^{2+} atoms. The white lines delineate a unit cell.

Radiative recombination at F^+ centres is not the only relevant process since ionizing radiation creates a host of other point defects in BeO that appear in its optical absorption and, as mentioned earlier, in its electrons spin resonance spectra [24, 53]. In particular, various types of hole centres that may influence radiative emission are possible. These include V^-, V^o and V_a type hole centres. The V^- describes a hole trapped at an O^{2-} cation vacancy in a nearest-neighbour position to a cation vacancy. The V_a centre arises when a V^- centre is associated with a triply-charged cation. Other notable types are those involving a hole trapped at an O^{2-} cation in the first neighbour position to a substitutional Na^+ or Li^+ impurity. The latter is specifically referred to as a $[Li]^0$ hole centre (e.g. [24, 27, 53]).

The diversity of point defects in BeO, some of which are radiation-induced, calls for a detailed explanation of the TL or OSL in terms of these imperfections. Indeed, various issues such as poor reproducibility of the dosimetric data in the low dose region [60] and inconsistencies in the TL and OSL models of BeO have beset its application to radiation dosimetry. One of these discrepancies relates to the electron trap whose TL peak is seemingly unresponsive to optical exposure and yet counterintuitively produces optically stimulated luminescence [61, 69]. Studies of the TL and OSL of BeO have therefore generated fewer answers to the many questions spawned. The same applies for its PTTL.

Some of the earliest studies on the PTTL of BeO [52, 61] created a perspective from which subsequent investigations inferred the conduct of PTTL experiments in this material. Crase and Gammage [61] and Tochlin *et al* [52] alluded to the emergence of PTTL in BeO preheated to 450 °C. The result implied that PTTL occurs due to phototransfer from deep electron traps. This introduced a paradigm over which the study of PTTL in BeO has mostly been viewed. Measurements intended to examine whether BeO can luminesce under phototransfer have thus mostly been made where the glow curve has been cleared of all or most of its glow peaks.

Although the TL and OSL of BeO have been widely studied, there are comparatively fewer studies on its PTTL. Tochlin *et al* [52] alluded to it, Bulur and Göksu [70] surveyed it and Bulur [71] explored it. The BeO studied by Bulur [71] had three glow peaks at 75, 220 and 340 °C for heating at 5 °C s^{-1} after beta irradiation to 2.5 Gy. The PTTL appeared at 220 °C and 340 °C for irradiation to 10 Gy, preheating to 400 °C and illumination for 1500 s. The phototransfer was presumed to originate from a deep electron trap between 400 °C and 625 °C. In an earlier report, Bulur and Göksu [70] reported PTTL at 220 °C and 340 °C from 12.8 Gy irradiated BeO after preheating to 400 °C and also attributed the PTTL to deep electron traps. For these reports, samples were illuminated using a 420–550 nm broadband emitting halogen lamp. Isik *et al* [72] carried out kinetic analysis of PTTL measured at cryogenic temperatures and as previously [52, 61, 70, 71], they also associated the PTTL with deep electron traps. The studies of Bulur and Göksu [70] and Isik *et al* [72] were made on Thermalox^{TM} 995, a specific type of commercially available BeO. This type shows three glow peaks below 400 °C for heating at 5 °C s^{-1} [27, 71].

The PTTL of BeO has now been studied in detail by Chithambo and Kalita [73]. It is that study which is used in this section as an exemplar of measurement and analysis of PTTL from this material. The phototransfer was induced by 470 nm blue light, detected between 250 nm and 390 nm and refers to irradiation to 10 Gy.

4.4.2 Conventional thermoluminescence

4.4.2.1 Glow curve

Figure 4.16 is a collection of glow curves measured from BeO after irradiation to 1 and 10 Gy beta irradiation. The glow curve corresponding to 1 Gy shows three nominal peaks at 46, 178 and 290 °C. On the other hand, after 10 Gy there are five at 50, 182, 283, 437 and 535 °C, which we denote I, II, III, … in that order. Thermal cleaning reveals a previously concealed peak in the rising edge of peak III (figure 4.16 inset) [73].

4.4.3 Preparatory tests for PTTL

4.4.3.1 Peaks I–V

Preliminary tests to determine which peaks are reproduced under phototransfer show only the first three (I, II and III). Peaks IV, V and IIIA (the puny one in the inset to figure 4.16) are not regenerated under phototransfer. Examples of glow curves measured following illumination for 10 s after preheating to 77 °C and 360 °C intended to remove peak I and peaks I–III, respectively, are shown in figure 4.17.

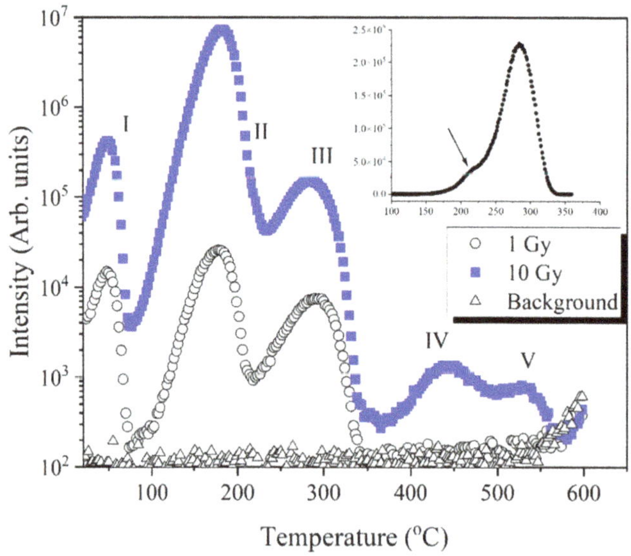

Figure 4.16. Increasing the dose from 1 Gy to 10 Gy not only changes the intensity of the TL from BeO but also introduces new peaks as is evident in the glow curves measured at 1 °C s^{-1} shown here. There is a weak peak in the rising edge of peak III (inset) which appears better when the preceding peaks have been glowed off. The material studied was BeO from Mirion Technologies (AWST) GmbH, Germany. Reprinted from [73], with the permission of AIP Publishing.

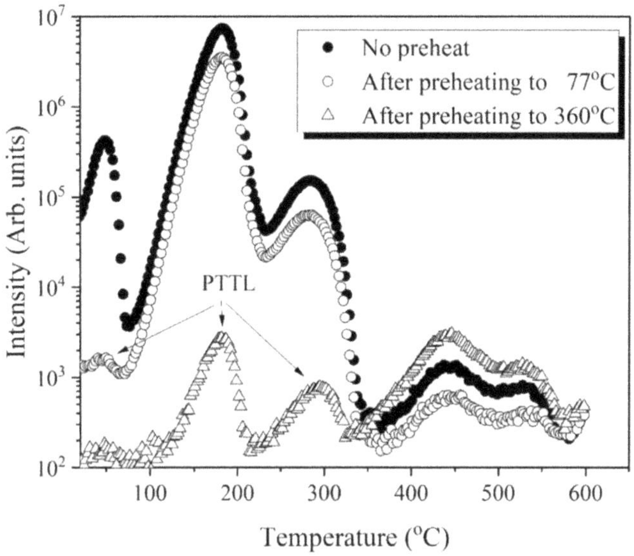

Figure 4.17. Glow curves measured after preheating that precede illumination shown alongside the original glow curve. The preheating removes peak I and peaks I–III, respectively. PTTL peaks are indicated. The original glow curve is included for comparison. Reprinted from [73], with the permission of AIP Publishing.

4.4.3.2 On phototransfer from deep electron traps

The longstanding perspective regarding BeO has been to associate its PTTL with only deep electron traps. Chithambo and Kalita [73] devised measurements to test this. In results that belie the view, Chithambo and Kalita [73] found that there is no PTTL in BeO preheated to 500 °C, illuminated for 10 s and reheated immediately thereafter. PTTL only appears when the illumination is lengthened but even so only at peaks II and III. This is shown in figure 4.18. Peak II is more intense than peak III when the BeO is illuminated for 100 s. This is reversed when the duration of illumination is increased to 1000 s with peak III becoming brighter. If the preheating temperature is changed from 500 °C to 600 °C, again no PTTL appears. Thus, one concludes that the electron trap(s) responsible for the PTTL seen at peaks II and III after preheating to 500 °C is/are activated between 500 °C and 600 °C. Of relevance is that Bulur and Göksu [70] and Bulur [71] attributed PTTL in BeO to donor electron traps between 500 °C and 600 °C. It is likely that the PTTL monitored at peak III originates from phototransfer from the electron trap of peak V [73]. Chithambo and Kalita [73] did not find experimental evidence of electron traps 'deeper than 600 °C' contributing to the PTTL. It has also been reported in the literature [61, 70] that heating BeO to 650 °C 'removes the PTTL effect'. Whether the assertion means that the PTTL effect is destroyed or that PTTL is absent after preheating to 650 °C but present otherwise is open to interpretation. On this, Chithambo and Kalita [73] found that the BeO produces PTTL despite annealing at 650 °C. The PTTL in question appears at peaks II and III (figure 4.18, inset) showing that the PTTL effect is never eradicated in BeO by any annealing.

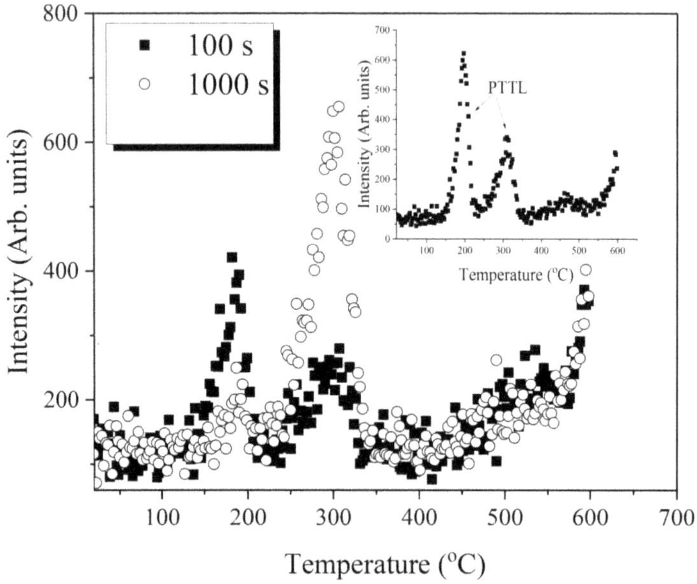

Figure 4.18. PTTL related to deep electron traps. Peak II is more intense than peak III for 100 s illumination. When the duration of illumination is lengthened to 1000 s, peak III becomes the brighter one. PTTL also results even after annealing at 650 °C for 30 min (inset). Reprinted from [73], with the permission of AIP Publishing.

4.4.4 Identifying donors and acceptors by pulse annealing

The contribution of electron traps as donors in the PTTL process can be assessed by monitoring the intensity of glow peaks as a function of preheating temperature. We exemplify this for BeO preheated from 80 °C to 650 °C at 10 °C intervals [73]. The interpretation of experimental behaviour is somewhat notional but is nevertheless useful as it underlies the mathematical analysis of the same data.

4.4.4.1 Phototransfer related to peaks I–V

The response of peaks I–III to preheating is shown in figure 4.19. We first look at peaks I–II in figure 4.19(a). These peaks weaken in intensity with preheating. The decrease of peak II between 180–200 °C occurs because its electron trap is depleted by preheating. Peak I is reproduced under PTTL in the same temperature range but gradually drops in intensity as the preheating temperature is increased. The concurrent decrease in intensity of peaks I and II suggests that the electron trap for peak II acts as a donor for phototransfer at peak I. There is thereafter minor change in intensity for both peaks because their putative donor is affected little by preheating. When preheating increases to 320 °C, during which peak III is removed, the PTTL at peaks I and II decreases sharply. These changes occur simultaneously, which points to the electron trap for peak III also serving as a donor for the PTTL at peaks I and II. When preheating extends beyond 320 °C, PTTL only appears at peak II but only if peak IV is present (figure 4.19(a), inset). The contribution of the electron trap for peak V to this PTTL must therefore be negligible. In summary, the

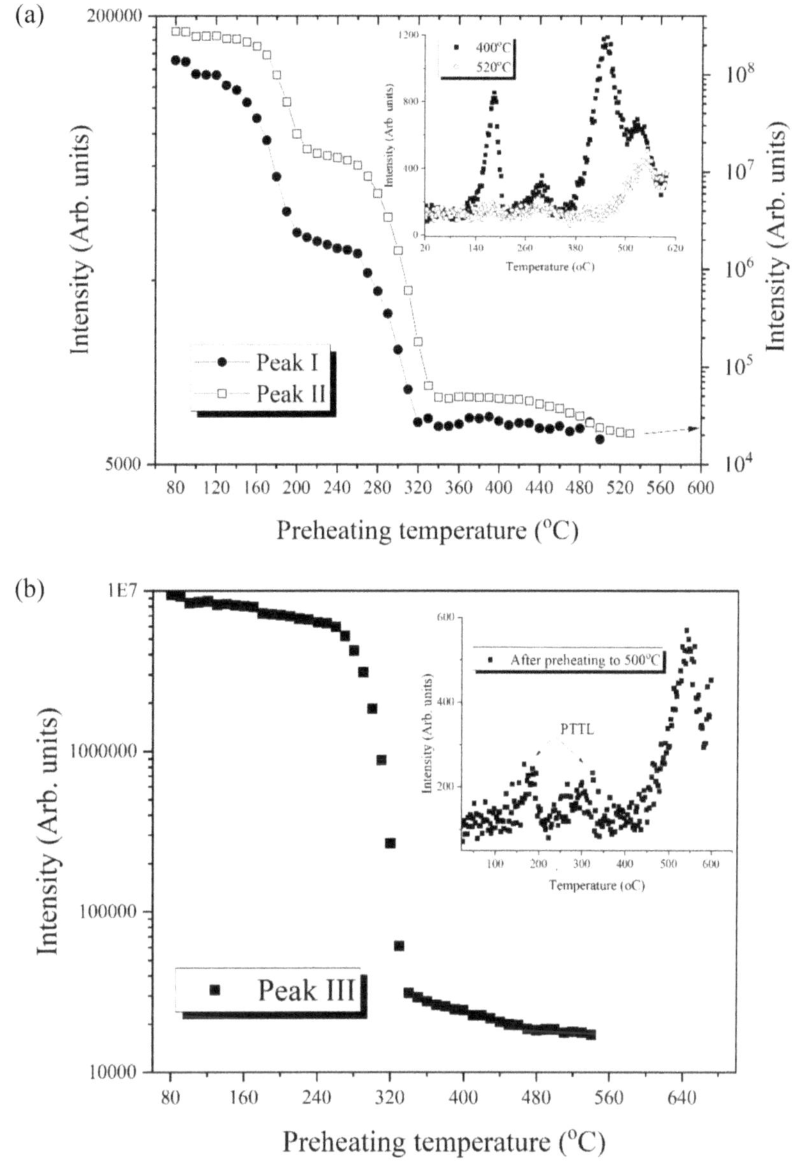

Figure 4.19. The effect of preheating temperature on peaks I and II (a) and III (b). The inset to part (a) shows measurements following preheating to 400 °C and 520 °C. The inset in part (b) is a glow curve measured after preheating to 500 °C. Reprinted from [73], with the permission of AIP Publishing.

PTTL at peak II occurs owing to phototransfer from the electron traps of peaks III and IV with the contribution from peak III being greater. In comparison, peak I is not reproduced under phototransfer for preheating over the same temperatures (figure 4.19(a), inset). All peaks are also light-sensitive and decrease with illumination [73].

Figure 4.19(b) displays the effect of preheating temperature on the intensity of peak III. Its intensity decreases between 80 °C and 260 °C since this peak acts as a donor for PTTL at peaks I and II over these temperatures. The loss of intensity that follows up to 340 °C is caused by the peak being progressively removed by preheating. Thereafter, peak III reappears under phototransfer but progressively fades since further preheating depletes its donors, namely, electron traps for peaks IV and V. Unlike peak II, peak III is still induced by phototransfer despite the removal of peak IV. The inset to figure 4.19(b) shows an example of a glow curve obtained after preheating to 500 °C with only peak V remaining. The presence of PTTL peaks II and III below 500 °C is obvious.

4.4.4.2 Peaks IV and V

Peaks IV and V are neither regenerated by phototransfer nor affected by preheating before the onset of their removal [73]. Their electron traps can only be weak donors for PTTL peaks I–III.

4.4.5 Thermally transferred luminescence

The models for PTTL in this work focus on the restorative processes caused by illumination. One links acceptors and donors if the change in their intensity is concomitant. This is not always the case. For completeness, we therefore deviate somewhat from the main objective of this chapter to instead look at recuperation or thermally transferred optically stimulated luminescence (TT-OSL) in BeO. This has been reviewed by Yukihara [69] for measurements on Thermalox^{TM} 995 type BeO, which has three stand-out peaks at about 75 °C (peak 1), 200 °C (peak 2) and 310 °C (peak 3) for heating at 5 °C s^{-1}.

In the pulse annealing measurements of Yukihara [74], the intensity change of the OSL resembles that of peak 3 and decreases when peak 3 also does. This is shown in figure 4.20. Since the changes are concurrent, once concludes that the electron trap of peak 3 is an important source of charge for the OSL and that its contribution exceeds that from peaks 1 and 2. This conclusion is, however, at odds with experiment, which shows that peak 3 is hardly affected by light.

Yukihara [74, 76] invoked the phenomenon of TT-OSL to address the discrepancy. In the model, TT-OSL is produced when some electrons thermally stimulated from the electron trap of peak 3 backscatter to an optically sensitive electron trap whose thermal stability is similar to and whose temperature position overlaps that of peak 3.

An important aspect of the model of Yukihara [69] is that TT-OSL is generated by preheating the sample to temperatures in the same range in which TL peak 3 becomes unstable. It is from this result that the conclusion was drawn that the electron trap associated with peak 3 may also be the origin of charge transferred to the electron traps responsible for both the OSL and TT-OSL [74]. The dependence of the TT-OSL intensity on duration of preheating and temperature is deduced to be consistent with a model with two types of electron traps (one optically responsive and the other, poorly responsive) that activate in a similar temperature range which is also isothermal to that of peak 3.

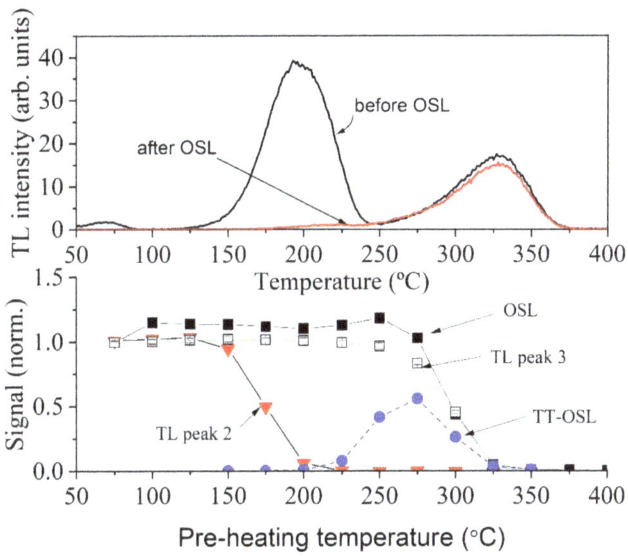

Figure 4.20. Glow curves of BeO obtained at 5 °C s^{-1} following beta irradiation to 200 mGy measured either before or after illumination with 470 nm blue light [75] (top panel) pulse annealing tests intended to compare the thermal stability of TL peaks 2 and 3, the resultant OSL and the induced TT-OSL [74] (bottom panel).

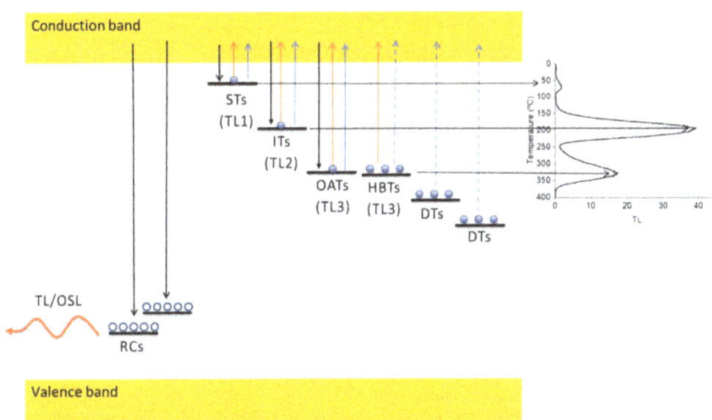

Figure 4.21. The model of Yukihara [74] to explain TT-OSL in BeO. The model describes the role of shallow traps (STs), intermediate-energy traps (ITs), optically active ones (OATs), quiescent types (HBTs), deep traps (DTs) and includes recombination centres (RCs) for completeness. Reprinted from [69], Copyright (2020), with permission from Elsevier.

A schematic diagram used by Yukihara [69] to account for TT-OSL in the BeO model is shown in figure 4.21. Ionizing radiation leads to filling of the electron trapping sites with electrons (solid symbols) and holes (open symbols). The various arrows represent thermal stimulation (orange arrows) and optical stimulation (blue ones). The dashed blue lines symbolise optical transitions associated with poorly light-sensitive or quiescent electron traps. Those are named hard-to-bleach traps (HBT) to distinguish them from others that are amenable to optical stimulation (referred to as optically active traps, OATs). A temperature-resolved plot of thermal stimulation for such a distribution of electron traps linking specific peaks to relevant electron traps is included.

Yukihara [69] explains TT-OSL with reference to figure 4.21. When the BeO is illuminated or irradiated, both quiescent and active electron traps are filled. During subsequent heating, some charge thermally released from the quiescent traps transits to depleted optically active traps. It is the thermally transferred charge that is responsible for the eponymously named thermally transferred OSL. Although our discussions do not specifically single out the role of the thermally transferred component in the PTTL in BeO, this is an interesting open question.

4.4.6 Time-response profiles

The combined effect of preheating and illumination is instructive in helping one to understand the role of donors in the PTTL process, and in some cases identifying the donors. The time-dependent behaviour of PTTL is never uniform and may also be used to explain mechanisms at play in the emission of the luminescence. In this part, we look at the influence of duration of illumination on the PTTL monitored at peaks I–III after preheating to remove peak I, peaks I and II, I–III and I–IV. Peaks I, II and III are the only three reproduced under phototransfer.

4.4.6.1 PTTL following removal of peak I only

When peak I is the only one removed by preheating, it reappears under phototransfer [73]. Its intensity rises to a maximum for short illumination before decreasing with longer light exposures.

4.4.6.2 PTTL following preheating that removes peaks I and II

When peaks I and II are cleared by preheating, they are both regenerated by phototransfer. Their PTTL arises due to phototransfer from electron traps of peaks unaffected by preheating, namely, peaks III–V. Figure 4.22 shows the time-response profiles of peaks I and II over 1000 s. The PTTL becomes increasingly intense in the initial transient but fades with further illumination.

4.4.6.3 PTTL following preheating that removes peaks I–III

When peaks I to III are removed by preheating, PTTL is only seen at peaks II and III. Figure 4.23(a) shows that the PTTL intensity of peak II passes through a maximum with illumination time. The inset compares two glow curves, one recorded after a brief illumination of 3 s and the other, after a much lengthier one, of 3000 s. Although no PTTL is observed at peak I, the PTTL that does materialise at peaks II and III is quite bright.

Figure 4.23(b) is a profile of PTTL intensity for peak III. The intensity of this peak increases throughout measurement as shown. Its increase in the first 3000 s or so is rapid but slows down thereafter. The putative donors for this PTTL are the electron traps of peaks IV and V. The corresponding change of their peak intensity is shown in the inset to figure 4.23(b). Although either peak fades with illumination, as

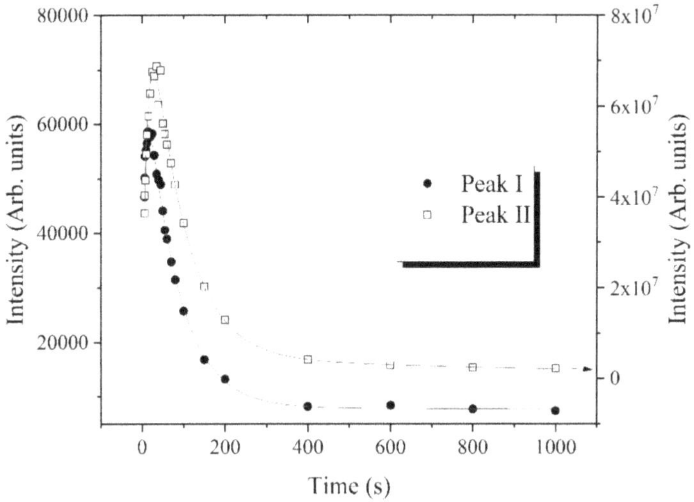

Figure 4.22. The dependence of PTTL intensity on duration of illumination for peaks I and II after preheating to deplete electron traps of peaks I and II. The lines through the data are mathematical results discussed in the text. Reprinted from [73], with the permission of AIP Publishing.

might be expected of a donor, the peaks are fainter than PTTL peaks II and III. This implies that their electron traps are only weak donors to the PTTL observed at peaks II and III. This is the same conclusion reached in the pulse annealing study. Such a disparity between the intensity of donor and acceptor is an interesting puzzle.

Since peaks IV and V are of low intensity, phototransfer from their electron traps is unlikely to be the only reason why appreciable PTTL is measured at peaks II and III. It can be rationalised that hole traps become unstable near 360 °C, the preheating temperature used to remove peaks II and III. We assume that when the BeO is heated this high, a large number of holes are released from hole traps and some transit to recombination centres. This would produce the same effect as increasing the concentration of recombination centres.

4.4.6.4 PTTL from deep traps

The interest in measurements where all peaks (I–III) are removed (in this case by heating to 500 °C) is that any resulting PTTL can be attributed to either electron transfer from the electron trap of peak V or deep electron traps. PTTL does appear after preheating to 500 °C but only when the illumination is lengthy. On the other hand, there is no PTTL when the sample has been heated to 600 °C. This restricts the choice of donors to the electron trap of peak V. For further discussion, we focus on change caused by the duration of illumination on the PTTL for peak III (figure 4.23(c)). Since only lengthy illumination induces PTTL, small increments in the duration of illumination beyond that which induces maximum signal are immaterial in effecting change in the PTTL intensity. This system then reduces to one of a single acceptor and a single weak donor.

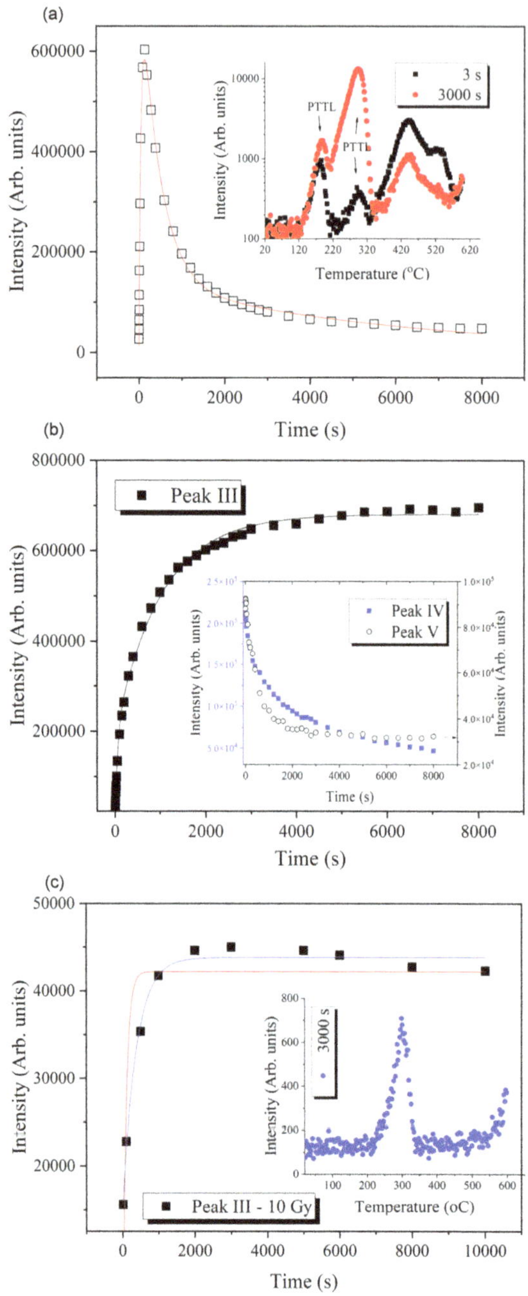

Figure 4.23. Time-response plots of PTTL peaks II (a) and III after preheating that removes the first three peaks (b) and for preheating to 500 °C where the donors are deep electron traps (c). The insets show that peak I is not regenerated under phototransfer. The lines through the data are touched on later in the text. Reprinted from [73], with the permission of AIP Publishing.

4.4.6.5 Influence of illumination on donor electron traps

Donor electron traps involved in PTTL are not equally important. Their role in phototransfer can be examined using their relative intensities for each preheating temperature. An example of how to do so is shown in figure 4.24. The figure displays, for a set of donors corresponding to each preheating temperature, a ratio plot of the most intense peak to all others. Thus, for instance, although PTTL following preheating that removes peak I only is associated with four possible donors, only two are of relevance. The resultant system can therefore be analysed in terms of a system of one acceptor and two donors. On the other hand, PTTL following preheating that clears peaks I and II can be associated with three donors whereas that resulting after the removal of the first three peaks, with two weak donors.

4.4.7 Analysis of PTTL time-response profiles

We now discuss the time-response profiles looked at qualitatively in the preceding section. We do this in three ways, namely, by a phenomenological model, a kinetics model and by way of vector fields.

4.4.7.1 Phenomenological model

Phenomenological analysis of PTTL intensity patterns is predicated on the fact that they can be analysed by considering electron transfer at the illumination stage only. This approach is developed and discussed in chapter 3. The particular solution giving the time dependence of the PTTL for a system of one acceptor and n donors is given in equation (3.18).

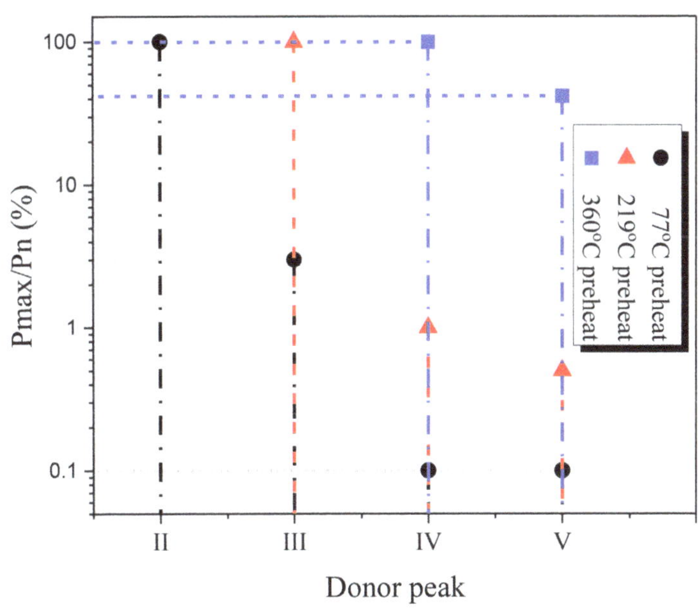

Figure 4.24. A ratio plot of the dominant donor to all other contributors in each set associated with a given preheating temperature. Reprinted from [73], with the permission of AIP Publishing.

Thus far we have relied on qualitative conclusions reached from both pulse annealing and notional interpretation of time-response profiles. It is interesting to see how well the phenomenological models match the experimental behaviour for PTTL that appears after removal of peaks I–II, I–III and I–V.

4.4.7.2 PTTL at peaks I and II after removal of peak I and peaks I–III

We recall that the PTTL induced after the removal of peak I only or peaks I–III corresponds to two donors. Using equation (3.18), the change of the PTTL with illumination time for peaks I, II and III can described as

$$N_a = B_1(e^{-f_a t} - e^{-f_1 t}) + B_2(e^{-f_a t} - e^{-f_2 t}) \tag{4.15}$$

where $B_1 = (\delta_1 f_1 N_{1i})/(f_1 - f_a)$ and $B_2 = (\delta_2 f_2 N_{2i})/(f_2 - f_a)$, where N_{1i} and N_{2i} are the initial concentrations at electron traps I and II. The result for peak II after removal of peaks I–III is shown fitted by equation (4.17) in figure 4.23(a).

4.4.7.3 Pattern of PTTL at peaks I and II after removal of peaks I–II

The PTTL at peaks I and II following removal of the first two peaks can in each case be considered as a system of one acceptor and three donors. One of these, electron trap peak V, is a minor contributor to the PTTL. The time dependence of the PTTL can also be described by equation (4.15) as figure 4.22 shows.

4.4.7.4 The differential impact of donors on PTTL at peak III

The form of peak III as phototransferred following preheating to 360 °C and 500 °C offers an interesting contrast between prediction of model and experimental result. When the sample is preheated to remove the first three peaks, i.e. to 360 °C in the study under discussion, the resultant system corresponds to two donors. On the other hand, when all peaks have been cleared by heating to 500 °C, the resultant PTTL can be associated with a single donor. Mathematical models based on either set-up turn out to be inconsistent with experimental results. The reason for the disagreement is that in formulating models, one assumes an ideal donor whose contribution to the phototransfer is meaningful. The experimental reality here is anything but and various approximations are necessary.

The notional interpretation of the PTTL process would suggest that the PTTL at peak III may also be fitted by equation (4.15). However, since of the two weak donors one can be neglected, one finds

$$N_a = A(e^{-f_1 t} - e^{-f_a t}) + B(1 - e^{-f_a t}) \tag{4.16}$$

where A and B are constants [73]. A fit of equation (4.16) to the PTTL time profile of peak III after preheating to 360 °C is shown in figure 4.23(b) and is a better description of the data.

When all peaks in the glow curve have been preheated off, the PTTL obtained at peak III after such preheating to 500 °C corresponds to the archetype single donor system. The light-induced loss of electrons from this weak donor, that is, the electron trap for peak V, can be approximated as $N_5 = N_{5i}e^{-f_5 t}$ where N_{5i} is its initial occupancy. The value of N_{5i} before any illumination is negligible, since $f_5 \ll 1$; $N_5 \approx N_{5i}$. If one re-casts the problem on the basis that the number of electrons optically transferred from the donor is negligible and that it remains so over time, we obtain the same result that $N_5 \approx N_{5i}$. With this simplification,

$$\frac{dN_3}{dt} = -f_3 N_3 + \gamma f_5 N_{5i} \tag{4.17}$$

from which

$$N_3 = \kappa(1 - e^{-f_3 t}) \tag{4.18}$$

where γ is a constant of proportionality and $\kappa = \gamma f_5 N_{5i}/f_3$. Equation (4.18) is a saturating exponential from the start of illumination at $t = 0$. On the other hand, if one retains the form $N_5 \approx N_{5i}(1 - f_5 t)$, the result obtained is

$$N_3 = A(1 - e^{-f_3 t}) - B(1 - Be^{-f_3 t}) - Ce^{-f_3 t} \tag{4.19}$$

where A, B and C are constants. The fit of equations (4.18) and (4.19) to the experimentally obtained PTTL profiles of peak III are compared in figure 4.23(c). We see that equation (4.19) (blue line) approximates the experimental behaviour slightly better.

4.4.8 Vector fields

Vector fields display the general shape of solutions of differential equations. A vector field offers a means to trace a trajectory of a solution in the phase plane, that is, the '*xy*' plane. Knowledge of the solution of the rate equation is not necessary to draw the vector fields. This method is thus a useful indicator of experimental behaviour. Figure 4.25 displays a vector field analogue of equation (4.17) where the inclination

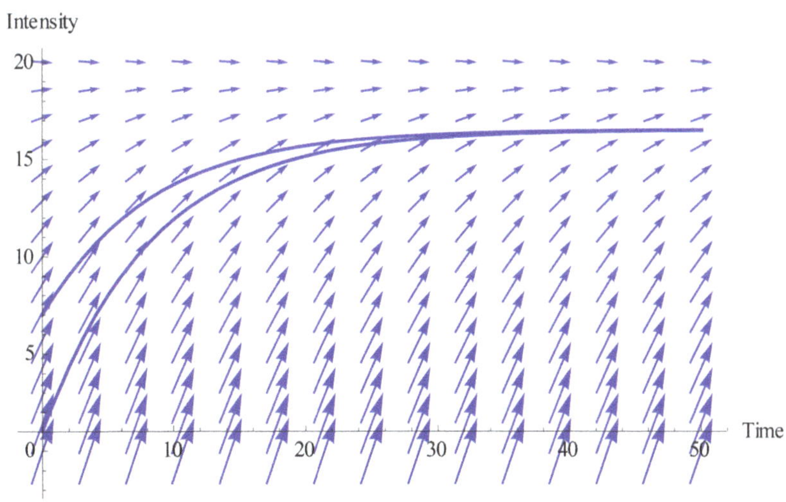

Figure 4.25. Vector field analogues of equation (4.17). Reprinted from [73], with the permission of AIP Publishing.

of the vector fields and trajectories of valid solutions are consistent with the experimental behaviour of figure 4.23(c).

4.4.9 Theoretical modelling

The time-dependent behaviour of PTTL can alternatively be treated by theoretically modelling the change of concentration at the electron traps by considering retrapping as well as the rate of electron transport through the conduction band. For the specific case of a system of three donors, the system of rate equations can be expressed as

$$\frac{dn_{d1}}{dt} = -f_{d1}n_{d1} + n_c(N_{d1} - n_{d1})A_{n1} \tag{4.20}$$

$$\frac{dn_{d2}}{dt} = -f_{d2}n_{d2} + n_c(N_{d2} - n_{d2})A_{n2} \tag{4.21}$$

$$\frac{dn_{d3}}{dt} = -f_{d3}n_{d3} + n_c(N_{d3} - n_{d3})A_{n3} \tag{4.22}$$

$$\frac{dn_c}{dt} = -A_m h n_c - \frac{dn_{d1}}{dt} - \frac{dn_{d2}}{dt} - \frac{dn_{d3}}{dt} \tag{4.23}$$

$$\frac{dh}{dt} = \frac{dn_{d1}}{dt} + \frac{dn_{d2}}{dt} + \frac{dn_{d3}}{dt} + \frac{dn_c}{dt} \tag{4.24}$$

$$I_{\text{PTTL}} = \frac{dh}{dt} = -A_m h n_c \tag{4.25}$$

where N_{di} is the concentration of the ith electron trap, n_{di} the instantaneous electron concentration at the ith electron trap, A_{di} and A_m are the retrapping and recombination probabilities, n_c and h are the instantaneous concentration of electrons in the conduction band and holes at the recombination centre and f_{di} the optical stimulation probability.

Figure 4.26 compares simulation (solid lines) with experiment (solid symbols) for peak I phototransferred after the removal of peaks I and II only (in part (a)). The PTTL here is related to three donors. Figure 4.26(b) is a simulation of the time-response profile for peak III owing to deep electron traps. These simulations favourably reproduce the experimental patterns. A list of systems of acceptor and donor(s) involved in the PTTL discussed is shown in table 4.2.

4.4.10 Mechanisms

It is helpful to consider the mechanisms for some aspects of PTTL in BeO. This sample discussed here has five peaks (I–V) of which only the first three are regenerated under phototransfer [73]. Although electron traps of peaks IV and V also act as donors, their contribution is minimal. It is therefore notable that PTTL at peaks II and III, supposedly due to these weak donors, is counterintuitively bright. Although our concern is PTTL, questions on the electronic processes responsible

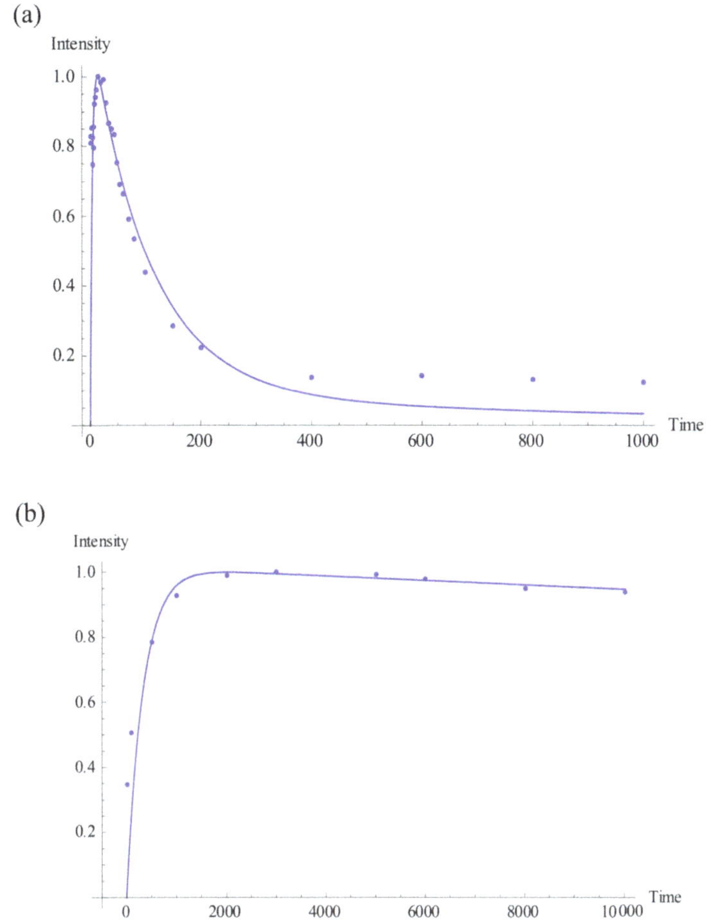

Figure 4.26. Comparison of theoretically simulated and experimentally obtained data for the time dependence of PTTL for peak I following heating that clears peaks I–II (a) for peak III after preheating to remove peaks I–III (b) and after preheating to 500 °C. The set of parameters used in the first simulation are $f_{d1} = 9.0 \times 10^{-3}$ s^{-1}, $f_{d2} = 6.8 \times 10^{-4}$ s^{-1}, $f_{d3} = 9.7 \times 10^{-4}$ s^{-1}, $N_{d1} = 10^{10}$ cm^{-3}, $N_{d2} = 10^{14}$cm^{-3}, $N_{d3} = 10^{12}$cm^{-3}, $A_{n1} = A_{n2} = A_{n3} = 10^{-15}$ cm^3 s^{-1}, $A_m = 10^{-6}$ cm^3 s^{-1}. The initial conditions were taken as $N_{d1}(0) = N_{d2}(0) = N_{d3}(0) = 10^{10}$ cm^{-3}; $n_c = 0$ cm^{-3}. The parameters used in part (b) are $f_{d1} = 7.0 \times 10^{-6}$ s^{-1}, $N_{d1} = 10^{18}$ cm^{-3}, $A_{n1} = A_{n2} = 10^{-23}$ cm^3 s^{-1}, $A_m = 300A_{n1}$ cm^3 s^{-1} with the initial conditions as $N_{d1}(0) = 10^{18}$ cm^{-3}; $n_c = 0$ cm^{-3}. Reprinted from [73], with the permission of AIP Publishing.

for stimulated luminescence in BeO and its applications, as explored elsewhere [27, 28, 69], are relevant.

There are a number of factors useful in understanding radiative emission in BeO. One of these concerns the structural similarity between BeO and other oxides of cubic or low symmetry as pointed out by Ogorodnikov and Kruzhalov [53]. The resemblance suggests that features of defects in BeO may be similar to those in oxides with which BeO shares features. The similarity between the 3.7 eV (335 nm) F^+ bands in BeO [68] and in Al_2O_3:C [27, 28, 43] is a case in point. In particular, the

Table 4.2. Systems of acceptor and donor(s) involved in the PTTL discussed. Reprinted from [73], with the permission of AIP Publishing.

Peak label	T_m (°C)	Pulse annealing Donors	Pulse annealing Main donor[1]	Preheating temperature (°C)	Time response Possible donors	Time response Main donors[2]	Qualitative summary
I	50	II III	II	77	II III IV V	II III	1A2D
				219	III IV V	III IV V	1A3D
II	182	III IV (weak?)	III	219	III IV V	III IV V	1A3D
				360	IV V	IV V	1A2D (1A 2 weak donors)
III	283	IV V	IV V	360	IV V	IV V	1A2D (1A 2 weak donors)
				500	V HT	V HT	1A1D (1A 1 weak donor)
IV	437	N/A	N/A	N/A	N/A		
V	535	N/A	N/A	N/A	N/A		

[1] From pulse annealing study.
[2] From time-response study.

electron traps involved in the stimulated luminescence of BeO may be certain substitutional impurities [77].

To start with, we recall that there is a considerable increase in the PTTL at peak III when the sample is preheated to deplete electron traps of the first three peaks (figure 4.23(b)). This result defies explanation based on electron traps only. A possible cause could be an increase in the concentration of holes when hole traps are thermally emptied when the sample is preheated in that way after irradiation. We invoke the mechanism for release of holes [66], namely,

$$V^{o} \xrightarrow{\text{heating}} V + h^{+}$$

to reason that this is the process by which holes are released by the preheating that precedes measurement of PTTL peak III. These free holes catalyse the formation of F^+ centres such that the intensity of PTTL at peak III exceeds that monitored when the preheating temperature is less than that at which the hole traps are thermally activated. Similar sensitivity increases, owing to thermal annealing of hole traps, have likewise been reported in Al_2O_3:C [35] and Al_2O_3:C,Mg [78].

A separate question is why no PTTL appears at peaks IV and V when, as known [77], heating beyond 360 °C (which in our case clears peaks I–III) also thermally anneals V^- and V_B hole traps. To address this disparity, we need to look at the form of various hole centres possible in BeO. Hole centres in binary oxides include V^-, V^o and V_a defects [53]. The V^- centre denotes a hole trapped at a cation vacancy. The V_a centre refers to the situation where a V^- centre is associated with a triply-charged cation. The creation of a V_a centre requires a cation vacancy to be set between a trapped hole and an F centre. The V^- centres are unlikely to be promoted to any significant extent since the process would require more F centres than are produced during synthesis. In the same way, the concentration of V_B centres should be low because their formation requires a cation impurity adjacent to a V^- type defect whose concentration is already low to start with. The lack of any PTTL at peaks IV and V must therefore be due to a negligible concentration of V_B and V^- hole centres.

4.5 Al_2O_3:Cr

4.5.1 Introduction

Oxide-based ceramics such as Al_2O_3:Cr [79], magnesium oxide [80] and composites of Al_2O_3-BeO [81–84] are of interest for high dose dosimetry. These materials supplement long established examples such as $CaNa_2(SO_4)_2$:Dy, $CaSO_4$:Dy [85] or LiF:Mg,Cu,P [86]. The latter are considered favourably owing to their linear dose response which extends up to kGy.

The utility of Al_2O_3:Cr ceramics for high dose dosimetry using x-rays [87–89], heavy charged particles (HCP) [90] or thermal neutrons [91] is well documented. The TL of Al_2O_3:Cr induced by x-ray or HCP irradiation is particularly bright. Shinsho *et al* [88] reported measurements where the TL intensity of Al_2O_3:Cr is fivefold that of the undoped variety. This high sensitivity is also apparent when the Al_2O_3:Cr is exposed to helium (11.5 keV μm^{-1}) and C beams (13.3 keV μm^{-1}) [92] before use. When measured at 0.1 °C s^{-1} after such exposure, the glow curve displays a peak near 300 °C. This peak is preferred for use in dosimetry because its dose response is linear under x-ray and charged particle irradiation.

Since Al_2O_3:Cr retains its main TL characteristics even after heavy irradiation, it is suitable for study using phototransfer. The PTTL of Al_2O_3:Cr was studied by Chithambo *et al* [93]. Given the interest in Al_2O_3:Cr as a high dose dosemeter, the study paid particular attention to PTTL from deep electron traps and explored their dosimetric features as well as the effect of illumination temperature on the PTTL intensity. The presentation here draws on that work, which is as yet, the only one on PTTL of Al_2O_3:Cr.

4.5.2 Glow curve

Figure 4.27 shows a glow curve of Al_2O_3:Cr measured up to 650 °C at 1 °C s^{-1} after beta irradiation to 162 Gy. There are at least seven glow peaks. Peak V at 340 °C stands out from the rest near 80, 125, 169, 239, 468 and 580 °C labelled as shown. An eighth peak is only partially evident. There is an additional weak peak at 490 °C in the rising edge of peak VII which only becomes evident when preceding glow peaks have been removed. The position of each peak is not affected by irradiation dose, a feature of first order kinetics.

4.5.3 Tests for PTTL

Before any study of PTTL, the theoretical expectation that any peak not removed by preheating may correspond to a donor needs to be tested. In such test measurements [93], an irradiated sample of Al_2O_3:Cr was preheated in turn to 91, 135, 195, 255, 395, 500, 600 and 650 °C and illuminated for 100 s after each preheat. These temperature limits ensured that each of peaks I–VIII were removed in succession. The PTTL peak appears at peak

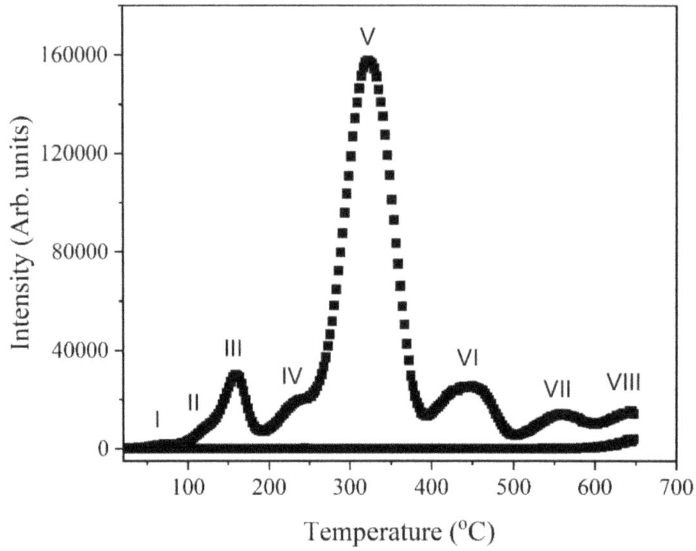

Figure 4.27. A glow curve of Al_2O_3:Cr measured at 1 °C s^{-1} following irradiation to 162 Gy. The background measurement is included for comparison. Reprinted from [93], Copyright (2023), with permission from Elsevier.

III, peaks III and IV, and peaks IV and V after preheating to 195, 255 and 500 °C in that order. There is no PTTL seen following removal of peak I, peaks I and II or all eight peaks for which the preheating temperatures are 91, 135 and 650 °C. All peaks except I, II and VIII reappear under phototransfer for the dose and illumination used. The PTTL following preheating to 395 °C and 600 °C, that is, after removal of peaks I–V, is weak and indistinct. At the outset, we can deduce that the electron trap of peak V is a major donor for PTTL and that any contribution from deep electron traps, if any, is negligible. Figure 4.28 shows an example of the PTTL measured after preheating to 500 °C (which removes peaks I–IV) and illumination for 100 s.

4.5.4 Pulse annealing

Measurements to qualitatively examine the role of electron traps as acceptors or donors used the pulse annealing routine on a sample irradiated to 162 Gy, preheated in turn between 50 °C and 650 °C at intervals of 10 °C and illuminated for 100 s each time. The changes in intensity of each of the peaks III–VII with respect to the preheating temperature are shown in figure 4.29. Using the same reasoning as before for the pulse annealing protocol, the PTTL at peak III can be linked to the depletion of electron trap V whereas the PTTL at peak V can be ascribed to phototransfer from electron traps of peaks VI and VII.

4.5.5 Influence of illumination time on PTTL

We consider time-response profiles, namely, the dependence of PTTL intensity on duration of illumination, after removal of peaks I–III, I–IV and I–V. The PTTL from deep electron traps refers to preheating up to 500 °C and 600 °C.

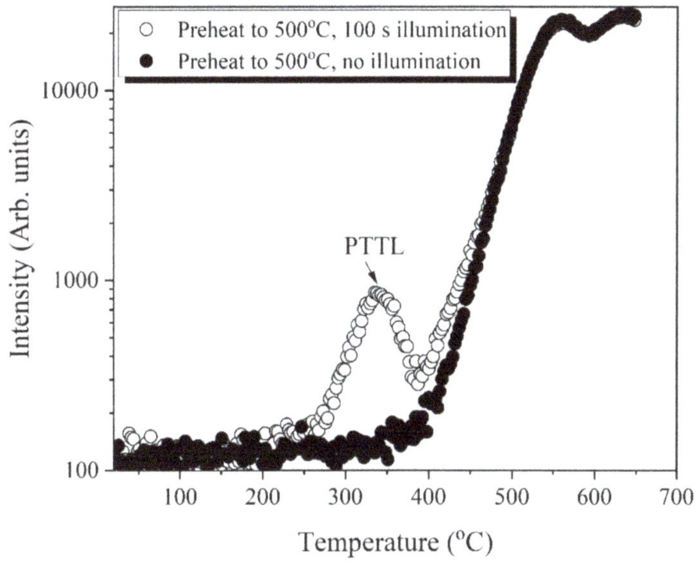

Figure 4.28. Glow curve measured following irradiation, preheating to 500 °C and illumination for 100 s. The resultant PTTL peak is clearly visible. A glow curve measured with similar settings except for illumination is included. Reprinted from [93], Copyright (2023), with permission from Elsevier.

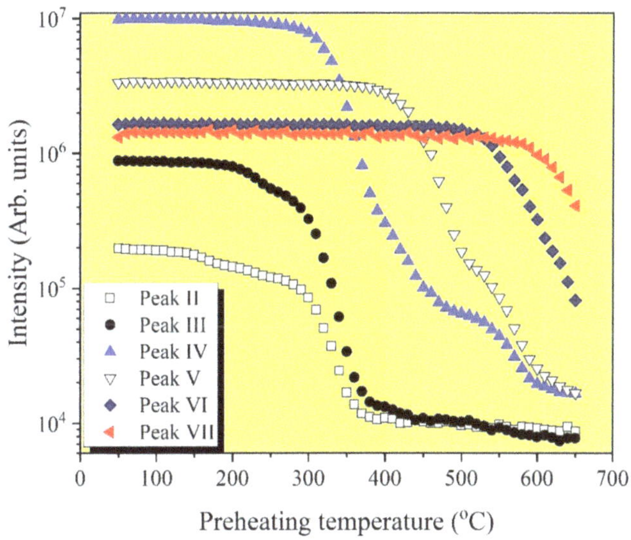

Figure 4.29. The effect of preheating on the intensity of peaks I–VII. Measurements were made at 10 °C intervals from 50 °C to 650 °C. Reprinted from [93], Copyright (2023), with permission from Elsevier.

4.5.5.1 PTTL following removal of peaks I–III

When the sample is preheated to remove peaks I–III, PTTL is only found at peak III. Its intensity decreases monotonically with illumination [93].

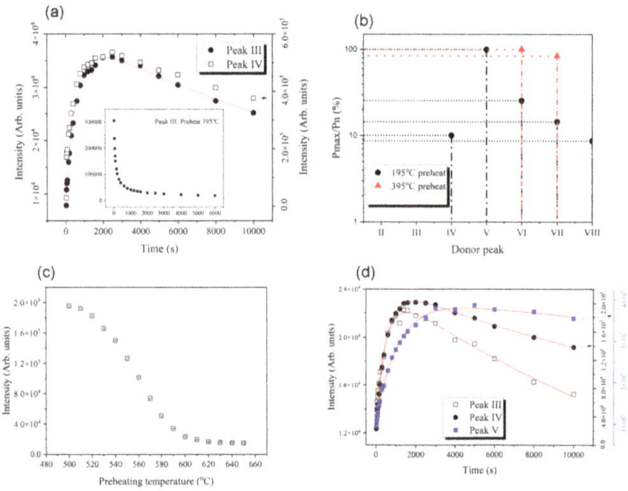

Figure 4.30. PTTL intensity patterns for peaks III and IV after preheating to 395 °C that clears peaks I–V and inset for peak III after preheating to 195 °C that removes peaks I–III (a). A ratio plot used to quantify the role of donors (b), application of pulse annealing for deep electron traps (c), PTTL time-response results of peaks III–V corresponding to preheating to 500 °C (d). Reprinted from [93], Copyright (2023), with permission from Elsevier.

4.5.5.2 PTTL following removal of peaks I–IV

If peaks I–IV are cleared prior to illumination, PTTL appears only at peaks III and IV. The time-response profiles of their PTTL also decreases consistently despite change in duration of illumination.

4.5.5.3 PTTL following removal of peaks I–V

Once peaks I–V have been removed by preheating, only peaks III and IV are reproduced by phototransfer. The influence of illumination on the intensity of these two peaks is shown in figure 4.30(a). The intensities go through a maximum and fall off thereafter.

4.5.5.4 Influence of illumination on donor peaks

The importance of specific donors in the PTTL can be deduced from figure 4.30(b). The plot shows that the PTTL measured after preheating to 195 °C to remove peaks I–III can be associated with two donors. Likewise, the PTTL following preheating to 395 °C to remove peaks I–V corresponds to two donors. A similar qualitative assessment is impractical for PTTL due to deep electron traps. However, one can quantify their contribution by monitoring the intensity of any resultant PTTL peak as the preheating temperature is increased. In this way, the intensity of peak V, the dominant one, is observed to decrease continually with preheating temperature (figure 4.30(c)). This is consistent with the expectation that the PTTL observed is from deep electron traps. A summary of systems of acceptor and donor(s) involved in the PTTL (including selected values of the photoionization cross section) is given in table 4.3.

Table 4.3. A summary of systems of acceptor and donor(s) involved in PTTL. Reprinted from [109], with the permission of AIP Publishing.

TL		PTTL					P. cross-section (cm^{-2})		
Peak label	T_m (°C)	Peak label	Preheating temperature (°C)	Possible donors	Main donor[1]	Main donor[2]	Acceptor	Donor	Donor
I	41		53	N/A	III	III	3.26E−18	3.92E−19	
II	80		101	III IV V VI VII	III	III IV	2.17E−18	2.17E−19	1.53E−20
III	110		127	IV V VI VII VIII	IV V VI	VII and VIII	5.25E−18	2.42E−19	4.70E−21
IV	140		180	N/A	N/A	N/A			
V	230		280		N/A	N/A			
VI	310		340		N/A	N/A			
VII	410		500		N/A	N/A			

[1] From pulse annealing study.
[2] From time-response study.

4.5.6 Properties of PTTL from deep electron traps

4.5.6.1 PTTL following preheating to 500 °C

When the sample is preheated to 500 °C to remove peaks I–VI, peaks III–V are reproduced under phototransfer. Their intensities gradually increase through a maximum (figure 4.30(d)).

4.5.6.2 PTTL following preheating to 600 °C

PTTL induced following preheating to 600 °C is intended to sense even deeper electron traps. PTTL is emitted near 330 °C and the change of intensity resembles that of the PTTL seen in figure 4.30(d) [93].

4.5.7 Analysis of PTTL time-response profiles

The mathematical analysis of time-response profiles is done using the same phenomenological models described in chapter 3. The main point is that if $f_i N_i$ is the number of electrons stimulated from the ith electron trap, the proportion scattered to an acceptor is $\delta_i f_i N_i$, where δ_i is a constant of proportionality. The time response of PTTL corresponding to preheating to 500 °C (figure 4.30(d)) is associated with at least two donors. The lines through the data here are, in each case, a fit of equation (3.18) with $j = 2$. The same model describes the PTTL following preheating to 395 °C for peaks III and IV (figure 4.30(a)). The first example, figure 4.30(d), emphasizes the role of deep electron traps because the only possible donors when the sample is heated to 500 °C are deep electron traps. The choice of two donors only means that this is the minimum number of terms required to produce a meaningful fit.

4.5.7.1 Vector fields and stability

We consider a vector field analogue of a system of one acceptor and one donor. This is also valid for a system of one acceptor and two donors if one of the latter makes a modest contribution to the PTTL. In our example, this may be relevant for peak V after preheating to 500 °C as shown in figure 4.30(d).

The phototransfer for a system of one acceptor and one donor may be expressed as

$$\begin{pmatrix} N_d' \\ N_k' \end{pmatrix} = \begin{pmatrix} -f_d & 0 \\ \delta_d f_d & -f_k \end{pmatrix} \begin{pmatrix} N_d \\ N_k \end{pmatrix} \tag{4.26}$$

where all symbols are as defined before. The critical point of equation (4.26) giving its equilibrium solution is (0,0). The eigenvalues of the matrix of coefficients are found as $-f_d$, and $-f_k$. Since these are both negative, the critical point should therefore be a stable improper node. Figure 4.31 shows the phase portrait of equation (4.26). The equilibrium solution is indeed stable and improper. If we consider only the physically meaningful first quadrant, we see that the portrait suggests that small changes in the initial electron concentration at the acceptor or donor will not cause any significant deviation from the equilibrium solution. If the

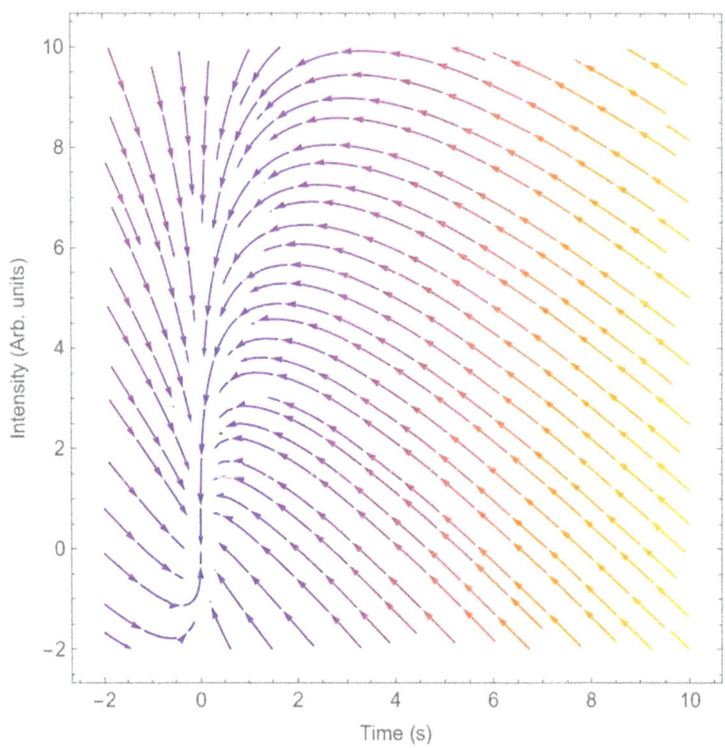

Figure 4.31. A vector field analogue of equation (4.26). Reprinted from [93], Copyright (2023), with permission from Elsevier.

duration of illumination is extensively long, the concentration at the acceptor electron trap increases to a maximum and then decreases and asymptotically approaches zero. Although slight changes in the initial conditions may affect the profiles shown, the resulting patterns will remain similar.

4.5.7.2 Theoretical modelling

If N_{di} is the concentration of the *i*th electron trap, n_{di} the instantaneous electron concentration at the *i*th electron trap, A_{di} and A_m are the retrapping and recombination probabilities, n_c and h the instantaneous concentration of electrons in the conduction band and holes at the recombination centre and f_{di} the optical stimulation probability, one can write for a system with two donors that

$$\frac{dn_{d1}}{dt} = -f_{d1}n_{d1} + n_c(N_{d1} - n_{d1})A_{n1} \tag{4.27}$$

$$\frac{dn_{d2}}{dt} = -f_{d2}n_{d2} + n_c(N_{d2} - n_{d2})A_{n2} \tag{4.28}$$

$$\frac{dn_c}{dt} = -A_m h n_c - \frac{dn_{d1}}{dt} - \frac{dn_{d2}}{dt} \tag{4.29}$$

$$\frac{dh}{dt} = \frac{dn_{d1}}{dt} + \frac{dn_{d2}}{dt} + \frac{dn_c}{dt} \tag{4.30}$$

$$I_{\text{PTTL}} = -\frac{dh}{dt} = A_m h n_c \tag{4.31}$$

Figure 4.32(a) shows the simulation of the time-response profile for peak IV corresponding to 395 °C, whereas figure 4.32(b) displays a simulation of the

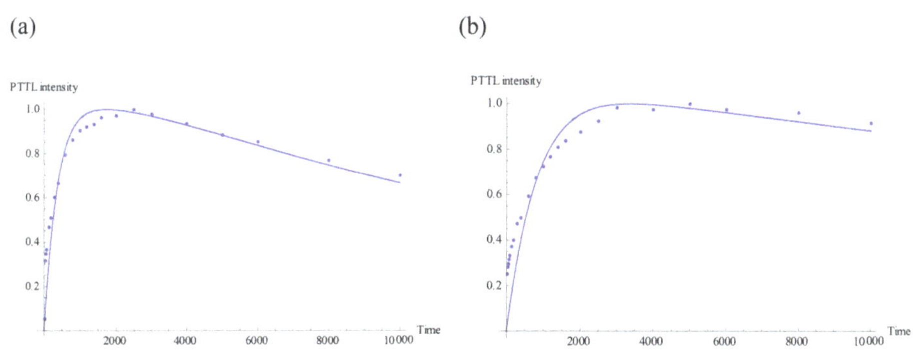

Figure 4.32. A comparison of simulated and experimental results for the time dependence of PTTL for peak IV following preheating that removes the first three peaks (a) for peak V after preheating to 500 °C (b). Reprinted from [93], Copyright (2023), with permission from Elsevier.

time-response profile for peak V following preheating to 500 °C. In both cases, the models are consistent with the experimental results.

4.5.8 Mechanisms

The luminescence properties of Al_2O_3 and its variants are well documented [24, 27, 28]. Al_2O_3 has emission bands at 1.25, 1.49, 3.0, 3.27 and 3.8 eV, that is, near 990, 830, 410, 380 and 330 nm. Much interest has been on the higher energy ones at 3.0, 3.27 and 3.8 eV since these offer the advantage of being well separated from infrared emissions from blackbody radiation that besets signal detection in this region, and also being coincident with detection bands of conventional luminescence instruments. As measurements on Al_2O_3:C and Al_2O_3:C, Mg show [43, 94, 95], the intensity of the latter bands are subject to annealing, dopant type and concentration. There are therefore parallels in the mechanisms involved in the emission of PTTL in Al_2O_3, Al_2O_3:C, Al_2O_3:Cr and Al_2O_3:C,Mg. We therefore postpone discussion of the mechanisms involved in the PTTL in Al_2O_3:Cr to the next section concerned with Al_2O_3:C,Mg.

4.6 Al_2O_3:C,Mg

4.6.1 Introduction

Al_2O_3:C,Mg is a polymorphic phase insulator developed for optical data storage [96] exploited for neutron dosimetry [97] and investigated for use as a radiation sensor under x-ray [98], beta- [99, 100], gamma- [101] and UV irradiation [102, 103]. Al_2O_3: C,Mg has an extensive number of glow peaks which runs up to nine depending on the annealing temperature [104–106]. All glow peaks are of first order kinetics [104, 105]. In particular, when Al_2O_3:C,Mg is annealed at 700 °C, its glow curve shows eight peaks but this number increases to nine when the annealing temperature is upped to 900 °C. In these two cases, only the first four peaks are regenerated under phototransfer [78, 107]. Some studies on Al_2O_3:C,Mg annealed at 1200 °C explored features of PTTL and TA-OSL in this material [106].

Annealing Al_2O_3:C, Mg changes its glow peak count and modifies its emission spectrum. This is exemplified by Al_2O_3:C, Mg annealed at 700 °C, which shows at least eight peaks. Annealing at 900 °C introduces a ninth [104]. The primary luminescence emission in Al_2O_3:C, Mg is at 410 nm with secondary side bands at 325 nm and 485 nm [106]. Annealing at 700 °C or 900 °C decreases the intensity of the 485 nm emission. The emission bands at 325, 410 and 485 nm are thought to originate from F^+, F and F_2^+ electron centres, respectively. F centre emission dominates in Al_2O_3:C [27] but if Al_2O_3 is double doped with C and Mg, the formation of F^+ centres is enhanced. When two adjacent F^+ centres aggregate to form an F_2^{2+} centre, the aggregate has been mooted to be responsible for yet another emission at 510 nm [97]. The decrease of the emission band at 485 nm noted earlier is deduced to result from an endothermic dissociation of F_2^+ centres when Al_2O_3:C, Mg is annealed at 700 °C or higher [108]. As such, it can be expected that the concentration of F^+ centres in Al_2O_3:C, Mg would increase, and with it, the F^+ emission. This is indeed observed [106].

Investigations on the radioluminescence (RL) of Al_2O_3:C,Mg [94, 98, 99] have looked at some less considered aspects of its emission spectrum. Trinidade and Jacobsohn [99] reported that both F and F^+ emissions suffer thermal quenching. Chithambo *et al* [94] also demonstrated, using Al_2O_3:C,Mg annealed at 1200 °C, that although both F and F^+ bands are affected by thermal quenching, the change for the 330 nm emission is atypical and, unlike the F band, cannot be readily analysed by the Mott–Seitz model.

The phototransferred thermoluminescence of Al_2O_3:C,Mg induced by 470 nm blue, 525 nm green and 870 nm infrared light was reported by Chithambo *et al* [109]. The conventional TL glow curve that they measured up to 600 °C comprises seven peaks (I–VII). Interestingly, only peak II is regenerated under phototransfer, and in exceptional cases, peak IV as well. Time-response profiles were analysed using phenomenological and kinetics models. The report includes studies devoted to the influence of illumination temperature on PTTL and demonstrated aspects related to thermal assistance and thermal quenching. The long-term nature of time-response profiles was analysed by stability theory.

We typify measurement and analysis of PTTL in Al_2O_3:C,Mg using this study [109] where the sample studied is Al_2O_3:C,Mg (Landauer, Inc; Oklahoma, USA) pre-annealed at 1200 °C before use. Annealing at this temperature changes its emission spectrum such that the 330 nm band becomes dominant [94, 108] and also alters the number of peaks in its glow curve. Glow curves were recorded at 1 °C s^{-1} and after beta irradiation to 3 Gy.

4.6.2 Spectral emission features

Characteristics of RL and TL spectra of unannealed and Al_2O_3:C,Mg annealed at 1200 °C are described elsewhere [109]. For measurements carried out at ambient temperature, RL is observed mainly at 410 nm for the unannealed sample but at 330 nm and 410 nm when Al_2O_3:C, Mg is annealed. These two bands stand out and depress weaker intensity features in the emission spectra. Figure 4.33 compares RL measured at 220 °C from annealed and unannealed material. Obtaining RL spectra at a temperature this high enables otherwise obscured bands at 330, 420, 510 and 700 nm to appear. Thus, one better sees the first three bands attributed to F^+, F and F_2^{2+} centres, respectively. The band at 700 nm, which is also observed in Cr-doped Al_2O_3 [110] and in transition-metal doped beryl [111] relates to transitions at Cr^{3+}, which homovalently substitutes for Al^{3+}.

4.6.3 Glow curve

Figure 4.34 is a glow curve measured at 1 °C s^{-1} after irradiation to 3 Gy. The plot is made over a semi-logarithmic scale to improve the dynamic range and thereby show peaks that might otherwise be concealed on a linear scale. There are at least seven nominal glow peaks. The main peak at 84 °C and subsidiary ones at 41, 110, 142, 230, 310 and 410 °C are labelled I–VII. Evidence of other electron traps has been noted as well [109]. Thermal cleaning turned up an additional ill-defined peak at 428 °C.

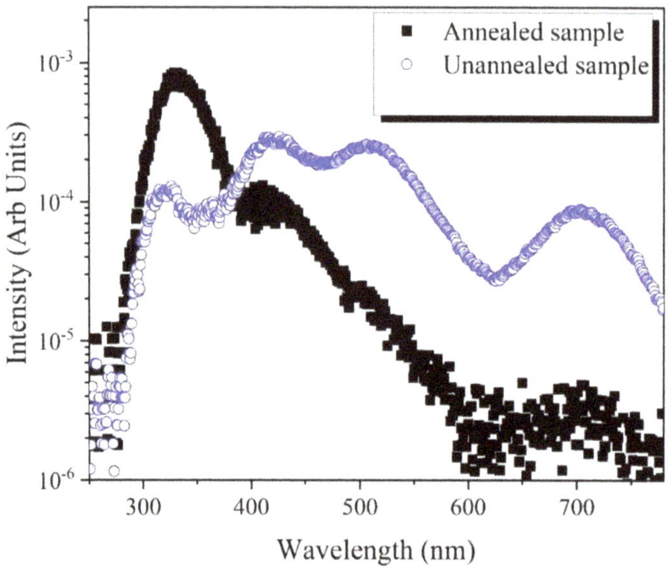

Figure 4.33. Radioluminescence spectra of unannealed and Al_2O_3:C,Mg annealed at 1200 °C. Both measurements were made at 220 °C to enable otherwise concealed secondary emission bands to appear. Reprinted from [109], with the permission of AIP Publishing. The spectral emission features were measured using RL in 1 mbar vacuum using a high sensitivity spectrometer [112, 113]. Samples were irradiated at a dose rate of 1.8 Gy min^{-1} with x-rays from a Philips MCN-101 x-ray tube set at 20 kV/4 mA. The luminescence is detected by a combination of two spectrometers capable of recording emissions over 250–850 nm.

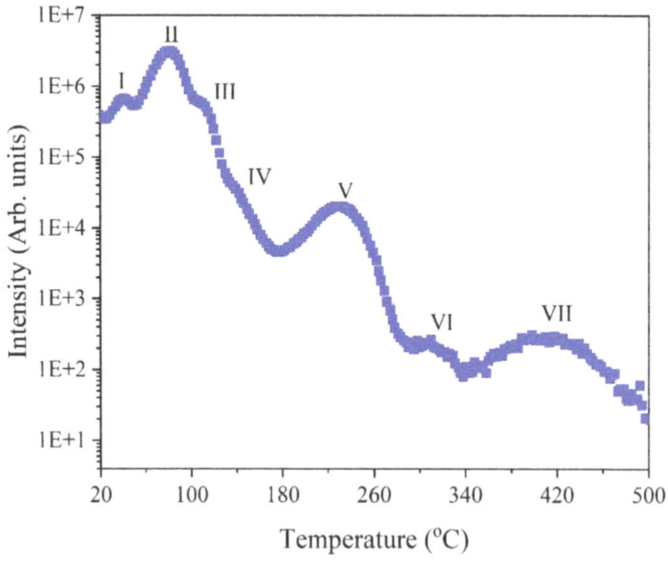

Figure 4.34. A glow curve of Al_2O_3:C,Mg after irradiation to 3 Gy. The sample was annealed at 1200 °C before use. The annealing not only changes its emission spectrum but also alters the number of glow peaks in the glow curve.

4.6.4 Tests preparatory to recording of PTTL

4.6.4.1 Wavelength of optical stimulation and PTTL

Figure 4.35 displays glow curves obtained following preheating to 280 °C and illumination for 10 s using 470 nm blue, 525 nm green and 870 nm LEDs. The preheating removed peaks I–V. PTTL appears only when Al_2O_3:C,Mg is illuminated by either blue or green light.

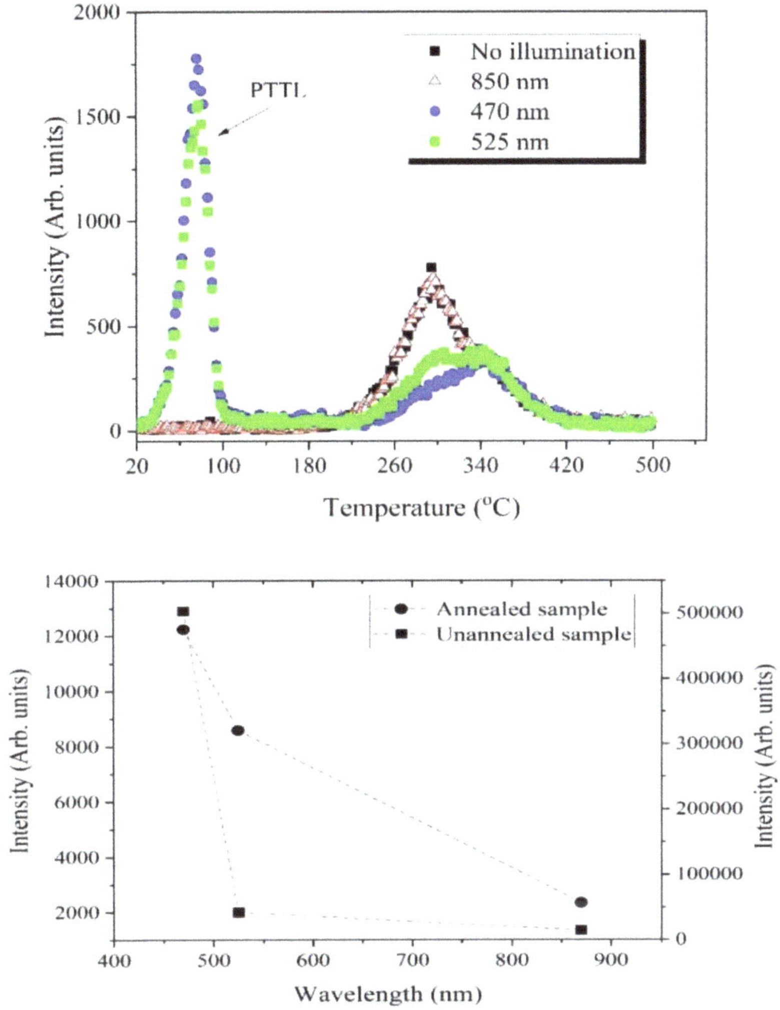

Figure 4.35. Glow curves measured after preheating to remove peaks I–V and illumination for 10 s by 470 nm blue, 525 nm green and 870 nm infrared LEDs (a). A glow curve recorded after preheating to 280 °C but without any illumination is included for comparison. The peak intensity against wavelength of illumination (b). Reprinted from [109], with the permission of AIP Publishing.

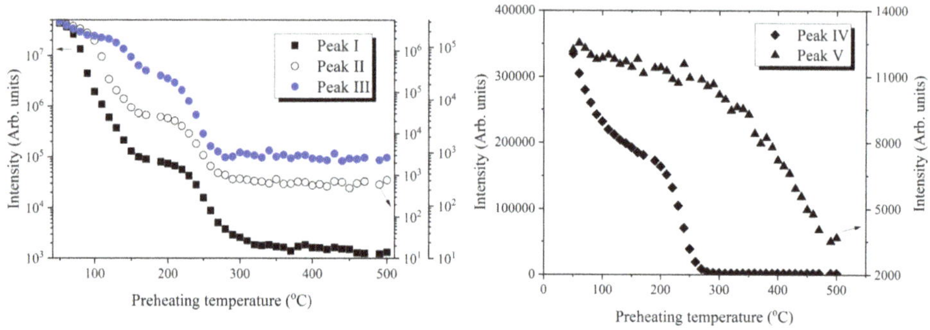

Figure 4.36. The effect of preheating temperature on peaks II, III and IV (a) and V and VI (b). Reprinted from [109], with the permission of AIP Publishing.

4.6.5 Pulse annealing

The role of electron traps as donors or acceptors in the PTTL is described using step annealing measurements from 50 °C to 500 °C at 10 °C intervals and 10 s illumination. Figure 4.36 shows the intensity against preheating temperature for peaks II, III and IV. The plot implies that the electron traps of peaks III–V are supposed donors for the PTTL observed at peak II. On the same principles for interpreting these types of plots, the PTTL at peak III is linked to the electron trap for peak V as a donor. Any PTTL produced at peak IV is weak and will mostly be due to phototransfer from the electron traps of peak VI and VII. Peaks V and VI are not regenerated by phototransfer [101].

4.6.6 PTTL time-response profiles

The dependence of PTTL intensity on duration of illumination aids in assessing the roles of acceptor and presumed donor electron traps. We consider the outcome of measurements corresponding to successive removal of glow peaks as a means to examine the influence of supposed electron traps on the acceptor.

4.6.6.1 PTTL after removal of peak I only

When only peak I is removed by preheating, no PTTL ensues regardless of any illumination.

4.6.6.2 PTTL after preheating to remove peaks I and II

PTTL only appears at peak II after preheating to remove peaks I and II. The influence of illumination on PTTL intensity is shown in figure 4.37(a). The result of measurements using green light illumination is included for comparison. In both examples, the intensity increases to a maximum then tails off. The emission induced by blue light peaks quicker and the rate of its decrease is greater. This is to be expected since blue light has comparatively greater energy. The possible donors for the PTTL observed at peak II are electron traps for peaks III–VII.

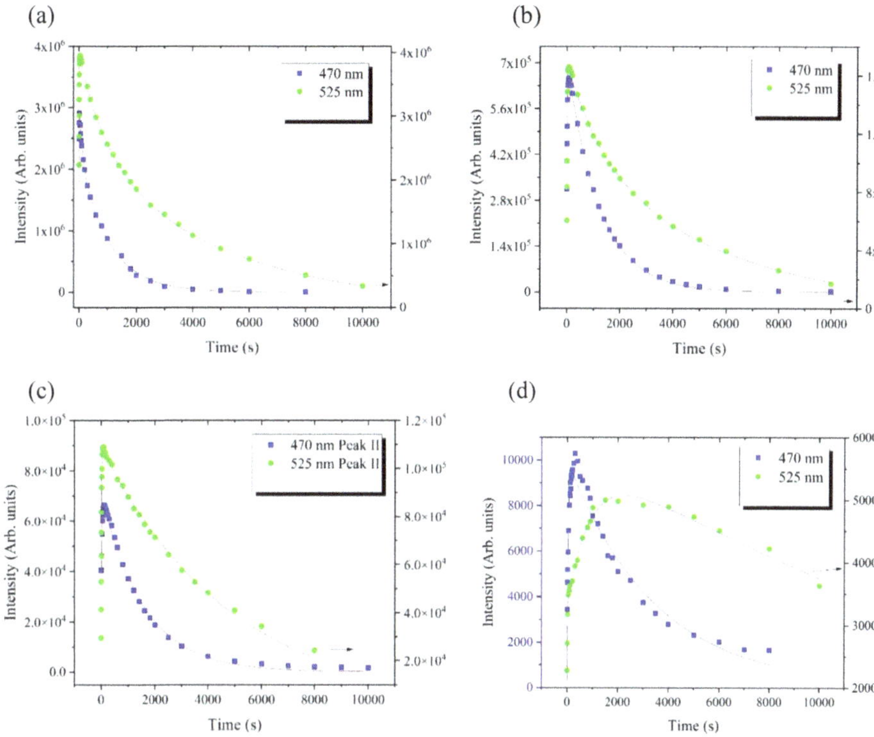

Figure 4.37. The time-dependent change of intensity induced under PTTL at peak II after blue or green light illumination corresponding to preheating to 101 °C to remove peak I (a) to 127 °C to remove peaks I and II (b) to 180 °C to clear peaks I–III (c) and 280 °C (d). The solid lines are fits based on equation (3.18). Reprinted from [109], with the permission of AIP Publishing.

4.6.6.3 PTTL after removal of peaks I–III

Once peaks I–III are removed by preheating, only peak II is reproduced under phototransfer. The electron traps of the remaining peaks, namely, IV–VII, are the possible donors for the PTTL. Figure 4.37(b) displays the time-response profile of the PTTL off peak II. Here, the peak also passes through a maximum under either blue or green light illumination.

4.6.6.4 PTTL after preheating to remove peaks I–IV

When the sample is preheated to remove peaks I–IV, peaks II and IV are the only ones that are reproduced under phototransfer. Figure 4.37(c) displays the PTTL time-response profiles of peak II where the possible donors are electron traps for peaks V–VII. The PTTL monitored at peak IV decreases continually with illumination [109].

4.6.6.5 PTTL after preheating to remove peaks I–IV

Following preheating to remove peaks I–IV and illumination, PTTL is found only at peak II. The influence of duration of illumination on the PTTL intensity is shown in figure 4.37(d). Whereas the intensity corresponding to blue light illumination goes through

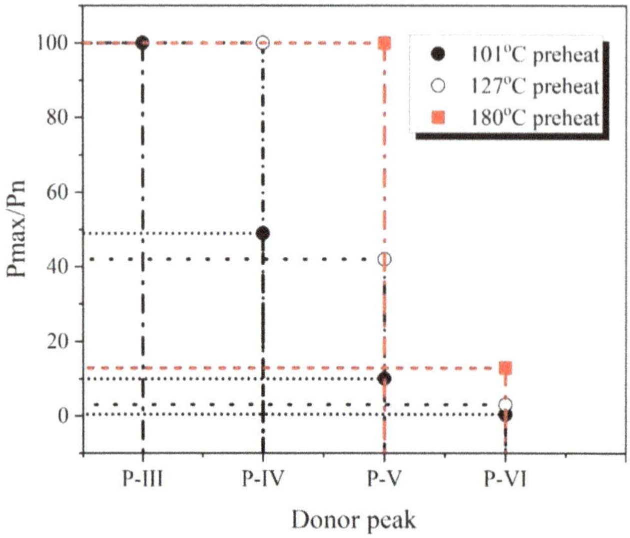

Figure 4.38. A ratio plot that denotes the degree of importance of each putative donor in the PTTL. The set of donors correspond to peaks III–VI and are determined by preheating to 101, 127 and 180 °C before illumination. These preheats remove the first two, the first three and the first four peaks in that order. Reprinted from [109], with the permission of AIP Publishing.

a maximum comparatively quicker, the change due to the green light is drawn out. The possible donors for the PTTL in this case correspond to electron traps of peaks VI and VII.

4.6.6.6 PTTL after preheating to remove peaks I–IV

When preheating removes peaks I–VI, there is essentially no PTTL monitored.

4.6.6.7 Effect of preheating to 500 °C and 600 °C

There is no PTTL following irradiation to 3 Gy used in the study and preheating to 500 °C or 600 °C.

4.6.6.8 Quantifying the role of donor electron traps

The contribution of putative donors can be quantified using the intensity of their corresponding peaks for each preheating. Figure 4.38 is a plot of ratios of the most prominent peak to all others in the set of donors associated with a particular acceptor. Here, the PTTL corresponding to preheating to 101 °C (which removes peaks I and II) is associated with four possible donors but we see that the contribution of the fourth donor at 0.4% is negligible. With similar reasoning, the PTTL corresponding to 180 °C (which removes peaks I–IV) corresponds to a single donor and that after 127 °C (which depletes electron traps of peaks I–III) to two donors.

4.6.7 Influence of stimulation temperature on PTTL intensity

Two factors that commonly affect the intensity of PTTL are the duration of illumination used to induce phototransfer and the temperature at which the material

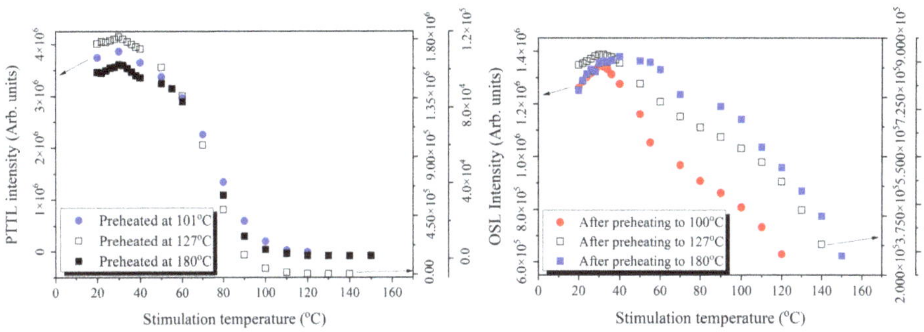

Figure 4.39. The influence of illumination temperature on the intensity of PTTL observed at peak II (left) and on the intensity of OSL recording during illumination (right). Measurements refer to preheating to 101 °C to remove peak I, to 127 °C to clear peaks I and II and to 180 °C to deplete the electron traps of the first three peaks. Reprinted from [109], with the permission of AIP Publishing.

is illuminated. The temperature-dependent changes for the PTTL and the attendant OSL for Al_2O_3:C,Mg are displayed in figures 4.39(left) and (right) respectively. Both sets of results, obtained after the first peak or the first two peaks or indeed the first three have been preheated off, are for PTTL at peak II. The intensities scale up owing to thermal assistance but not indefinitely because at some temperature, thermal quenching sets in.

4.6.8 Phototransfer from deep electron traps

We have used preheating to control the number of donors for PTTL. In order to restrict the source of the phototransfer to deep electron traps, one preheats to say, 500 °C or whatever heating limit an experimental rig allows. In practice, signals from deep electron trap are poor. To counter this drawback, it is necessary to considerably increase the dose and to irradiate the sample not at room temperature but at a temperature that ensures that only deep electron traps are filled (see figure 2.7).

4.6.8.1 The influence of thermal assistance and thermal quenching on phototransferred luminescence from deep electron traps

When an irradiated material is exposed to light of certain wavelengths, luminescence is optically stimulated. If the temperature at which the illumination is carried out is changed, the intensity of the luminescence changes too. Depending on the choice of temperature, the optically stimulated luminescence can either exceed or decrease below that measured at room temperature. Thermal assistance is usually cited to explain the increase since it is associated with the contribution of phonons to the photon-mediated detrapping of electrons. On the other hand, incidences of non-radiative recombination or thermal quenching usually underlie the temperature-related diminishing of luminescence intensity. Table 4.4 summarises three methodologies tailored to study these effects using luminescence from deep electron traps in Al_2O_3:C,Mg [109].

Table 4.4. Methods for experimental investigation of thermal assistance and thermal quenching using phototransfer from deep electron traps.

Method	Step	Procedure
1	1	Irradiate to 40 Gy at 500 °C
	2	Illuminate at intervals from 20 °C to 300 °C
2	1	Irradiate to 40 Gy at 500 °C
	2	Illuminate repeatedly at 20 °C only
3	1	Irradiate to 40 Gy at 500 °C
	2	Illuminate at temperature T
	3	Heat to 500 °C
	4	Repeat run but change the temperature in step 2

In the first method, the sample is irradiated to a high dose (40 Gy in this example) only once at the start of the experiment. The material is irradiated at 500 °C to supplant any need for preheating. There is no further irradiation between illumination in step 2. Although our concern is the component of optically stimulated luminescence due to temperature increase alone, it is in reality weak and easily obscured by signals due to optical stimulation alone. To circumvent this drawback, the LED power is reduced significantly. In this exemplar [109], the LED power was decreased to 20% (14 mW cm^{-2} at sample position) and the sample illuminated only for a brief 10 s each time. The control is a comparative measurement (method 2) in which the illumination is carried out only at room temperature.

The effect of illumination temperature and repetitive illumination on the luminescence optically stimulated from deep electron traps is compared in figure 4.40. In both cases, the intensity initially decreases consistently. The luminescence recorded with the temperature increasing is more than that monitored at ambient temperature only owing to thermal assistance to optical stimulation from deep electron traps. When the illumination temperature goes beyond 100 °C, the intensity of the luminescence measured at increasing temperatures drops below that obtained at room temperature due to thermal quenching.

4.6.8.2 Quantitative analysis of thermal assistance and thermal quenching

The methods described above can be adapted to serve as a means to quantify the effect of thermal assistance and thermal quenching on the luminescence from deep electron traps. The modification is summarised as method 3. To verify that any changes observed can be attributed to these effects only, measurements can be measured from say, 20 °C to 300 °C and repeated from 300 °C to 20 °C. These temperatures are particular to Al_2O_3:C.Mg [109]. Since thermal assistance and thermal quenching are independent thermal effects, the two sets can be used to verify that any changes observed can be attributed only to these two processes. The dependence of intensity on illumination temperature for such measurements on Al_2O_3:C.Mg [109] is shown in figure 4.41. The

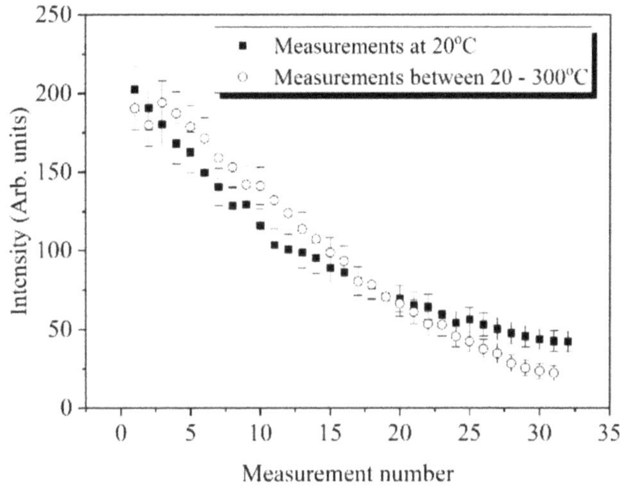

Figure 4.40. Change in intensity of luminescence recorded from Al_2O_3:C,Mg under simultaneous heating and optical stimulation (open symbols) compared to that monitored under optical stimulation only (solid symbols). Reprinted from [109], with the permission of AIP Publishing.

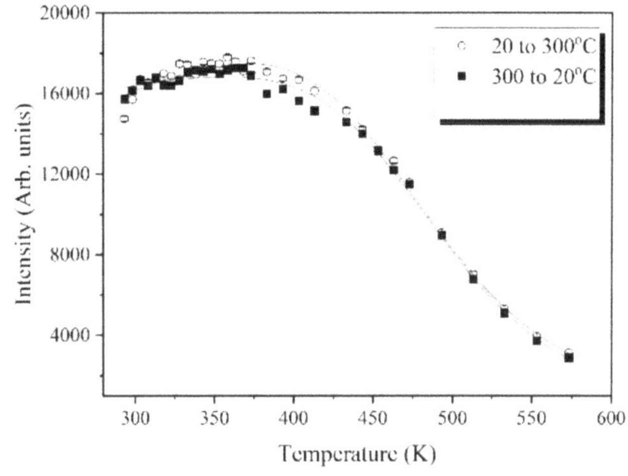

Figure 4.41. The dependence on measurement temperature between 20 °C and 300 °C of thermally-assisted optically stimulated luminescence from deep electron traps. The sample was irradiated before each illumination using 470 nm blue light at temperatures between 20 °C and 300 °C. Reprinted from [109], with the permission of AIP Publishing.

intensity patterns resemble and are indeed independent of the direction of temperature change.

Alternatively, instead of reducing the power of a lower wavelength light source, one can replace it a higher wavelength (and so weaker power) light source. Figure 4.42 shows results obtained with measurements made using 525 nm green

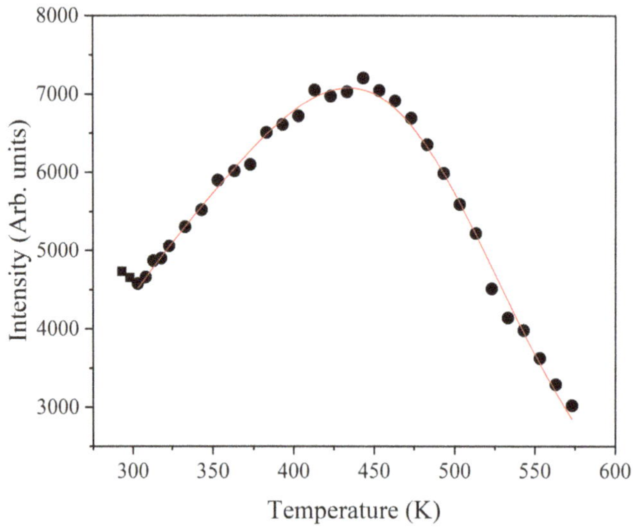

Figure 4.42. When Al_2O_3:C,Mg is irradiated at 500 °C between illumination at different temperatures using 525 nm LED light, the temperature-resolved plot of intensities goes through a peak as demonstrated here [109]. Reprinted from [109], with the permission of AIP Publishing.

Table 4.5. Values of the activation energy for thermal assistance and thermal quenching related to OSL and phototransfer from deep electron traps [109].

Wavelength (nm)	E_a (eV)	E_q (eV)	References
470	0.015 ± 0.002	0.47 ± 0.02	Figure 4.41/open symbols
470	0.010 ± 0.002	0.51 ± 0.02	Figure 4.41/ solid symbols
525	0.047 ± 0.002	1.01 ± 0.12	Figure 4.42
525	0.051 ± 0.02	0.75 ± 0.07	Figure 4.42

light (at 90% power setting)—instead of 470 nm blue light (at 20% setting power setting). The intensity changes due to differences in illumination temperature stand out better and clearly do not depend on whether the luminescence is obtained with the temperature increasing or decreasing. Since the OSL proceeds during phototransfer, we presume a direct relationship between changes in the intensity of OSL and PTTL. Values of the activation energy of thermal assistance and thermal quenching found using the methods described in section 3.8.1 (see also reference [114]) are given in table 4.5.

4.6.9 The analysis of PTTL time-response profiles

4.6.9.1 Phenomenological model

The systems for the PTTL measured at peak II from Al_2O_3 :C, Mg after peaks I, I–II or I–III are removed consist of three donors, two donors or a single donor. Plots for the dependence of PTTL intensity on illumination time shown in figure 4.37 are consistent with this expectation as analysed by equation (3.18). Figure 4.37(d) shows two interesting exceptions which, despite corresponding to three donors, are well described by use of a single donor. The caution here is that not all assigned donors make a material contribution to the PTTL.

4.6.9.2 Computational modelling

We also treat the time dependence of PTTL using kinetics-based modelling. The transport of electrons for a system of one acceptor and j donors is

$$\frac{dn_{d1}}{dt} = -f_{d1}n_{d1} + n_c(N_{d1} - n_{d1})A_{d1} \tag{4.32}$$

$$\frac{dn_{dj}}{dt} = -f_{dj}n_{dj} + n_c\left(N_{dj} - n_{dj}\right)A_{dj} \tag{4.33}$$

$$\frac{dn_c}{dt} = -A_m m n_c + \sum_i^j -f_{di}n_{di} + n_c(N_{di} - n_{di})A_{di} \tag{4.34}$$

$$I_{\text{PTTL}} = -\frac{dm}{dt} = A_m m n_c \tag{4.35}$$

where N_{di} is the concentration of the ith electron trap, n_{di} the electron concentration at the ith electron trap, A_{di} the corresponding recombination probability and A_m the recombination probability at the recombination centre. Other symbols retain their usual meanings. The set of j equations describe the optical stimulation and retrapping of electrons at donor electron traps. Figure 4.43 shows simulations (solid lines) overlaid with experimentally measured PTTL intensity values for peak II after preheating to remove peak I (left) or peaks I and II (right). The image on the left

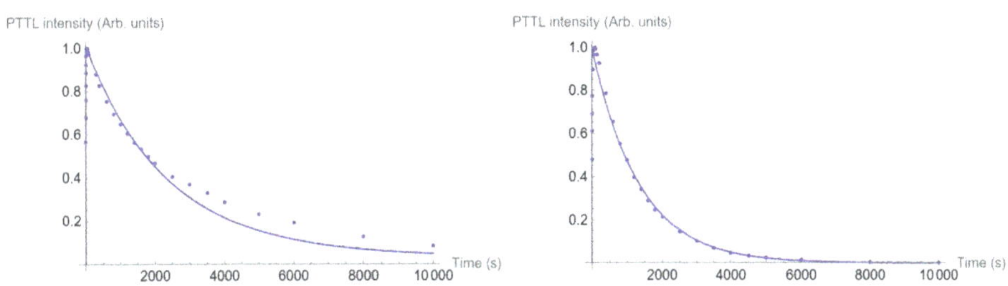

Figure 4.43. The change with illumination time of PTTL obtained after preheating to remove either the first peak only (left) or the first two peaks (right). The parameters used in the model are $f_{d1} = 4.0 \times 10^{-4}$ s^{-1}, $f_{d2} = 1.0 \times 10^{-5}$ s^{-1}, $f_{d3} = 1.8 \times 10^{-5}$ s^{-1}, $N_{d1} = 10^{11}$ cm^{-3}, $N_{d2} = 10^{14}$cm^{-3}, $N_{d3} = 10^{10}$cm^{-3}, $A_{d1} = A_{d2} = A_{d3} = 10^{-6}$ cm^3 s^{-1}, $A_m = 10^{-11}$ cm^3 s^{-1}. The initial conditions were taken as $N_{d1}(0) = N_{d2}(0) = N_{d3}(0) = 10^{10}$ cm^{-3}; $n_c = 0$ cm^{-3}. For the right image, the final set of parameters are $f_{d1} = 9.0 \times 10^{-4}$ s^{-1}, $f_{d2} = 8.0 \times 10^{-4}$ s^{-1}, $N_{d1} = 10^{10}$ cm^{-3}, $N_{d2} = 10^{11}$cm^{-3}, $A_{d1} = 10^{-6}$ cm^3 s^{-1}, $A_m = 10^{-11}$ cm^3 s^{-1}. Reprinted from [109], with the permission of AIP Publishing.

corresponds to three donors whereas this number decreases to one for the image on the right (figure 4.43).

4.6.10 Mechanisms

The mechanism for PTTL in Al_2O_3:C,Mg can, in outline, be understood in similar terms as for Al_2O_3:C, with which it shares major physical features. The puzzling experimental result is the fact that in Al_2O_3:C,Mg annealed at 1200 °C, PTTL is produced only at peak II. Although a tempting conclusion is that the concentration of its electron traps are dominant, it may also be that hole traps are involved in the PTTL process since the presence of a hole trap influencing the glow curve of Al_2O_3:C,Mg has been observed [115]. This poses yet another question of whether annealing affects the concentration of hole traps and, if it does, how that influences the PTTL. These and other questions are waiting for answers.

4.7 Summary

We have presented and discussed the measurement and analysis of phototransferred thermoluminescence from selected luminescent synthetic materials. Where possible, we have also attempted to explain mechanisms at play. The number of synthetic materials of interest is considerable as many reviews attest (e.g. [27, 116]). Although this is matched by the body of work on their synthesis, TL and OSL, and applications, the number of studies devoted to their PTTL is meagre. The understanding of this process in the few materials that have been investigated is understandably nascent. The examples in this chapter are not meant to be exhaustive and were chosen owing to ease of access to data. Early on, with reference to a certain material, we observed that studies of several phenomena including its phototransferred thermoluminescence has spawned more questions than answers. It is not by design that the penultimate section of this chapter ended with questions.

References

[1] Kristianpoller N, Abu-Rayya M and Chen R 1993 Phototransfer studies in synthetic quartz *Radiat. Prot. Dosim.* **1–4** 37–40

[2] Alexander C S, Morris M F and McKeever S W S 1997 The time and wavelength response of phototransferred thermoluminescence in natural and synthetic quartz *Radiat. Meas.* **27** 153–9

[3] Kombe-Atang E F M and Chithambo M L 2016 Phototransferred thermoluminescence and phosphorescence related to phototransfer in annealed synthetic quartz *Proc. of the 60th Annual Conf. of the South Africa Institute of Physics Conf.* pp 49–54

[4] Chithambo M L, Niyonzima P and Kalita J M 2018 Phototransferred thermoluminescence of synthetic quartz: analysis of illumination-time response curves *J. Lumin.* **198** 146–54

[5] Chithambo M L and Dawam R R 2020 Phototransferred thermoluminescence of annealed synthetic quartz: analysis of illumination-time profiles, kinetics and competition effects *Radiat. Meas.* **131** 106236

[6] Wintle A G and Murray A S 1997 Radiation measurements *Radiat. Meas.* **27** 611–24

[7] Santos A J J, de Lima J F and Valerio M E G 2001 Phototransferred thermoluminescence and phosphorescence related to phototransfer in annealed synthetic quartz *Radiat. Meas.* **33** 427–30

[8] Franklin A D 1997 On the interaction between the rapidly and slowly bleaching peaks in the TL glow curves of quartz *J. Lumin.* **75** 71–6

[9] Bertucci M, Veronese I and Cantone M C 2011 Photo-transferred thermoluminescence from deep traps in quartz *Radiat. Meas.* **46** 588–90

[10] de Souza L B F, Guzzo P L and Khoury H J 2014 OSL and photo-transferred TL of quartz single crystals sensitized by high-dose of gamma-radiation and moderate heat-treatments *Appl. Radiat. Iso.* **94** 93–100

[11] Bøtter-Jensen L, McKeever S W S and Wintle A G 2003 *Optically Stimulated Luminescence Dosimetry* (Amsterdam: Elsevier)

[12] Bailey R M 2000 The slow component of quartz optically stimulated luminescence *Radiat. Meas.* **32** 233–46

[13] Chithambo M L and Galloway R B 2001 On the slow component of luminescence stimulated from quartz by pulsed blue light-emitting diodes *Nucl. Instrum. Meth. B.* **183** 358–68

[14] Kitis G, Kiyak N G, Polymeris G S and Pagonis V 2010 On the feasibility of dating portable paintings: preliminary luminescence measurements on ground layer materials *Nucl. Instrum. Meth.* B **268** 592–8

[15] Gribble C D 1988 *Rutley's Elements of Mineralogy* (London: Unwin Hyman Ltd)

[16] Galloway R B 2002 Luminescence lifetimes in quartz: dependence on annealing temperature prior to beta irradiation *Radiat. Meas.* **35** 67–77

[17] Chithambo M L, Sane P and Tuomisto F 2011 Positron and luminescence lifetimes in annealed synthetic quartz *Radiat. Meas.* **46** 310–8

[18] Chithambo M L 2018 *An Introduction to Time-Resolved Optically Stimulated Luminescence* (Bristol: Morgan & Claypool Publishers)

[19] Chithambo M L and Niyonzima P 2017 Radioluminescence of annealed synthetic quartz *Radiat. Meas.* **106** 35–9

[20] Pagonis V, Chithambo M L, Chen R, Chruścińska A, Fasoli M, Li S H, Martini M and Ramseyer K 2014 Thermal dependence of luminescence lifetimes and radioluminescence in quartz *J. Lumin.* **145** 38–48

[21] Duller G A 1994 A new method for the analysis of infrared stimulated luminescence data from potassium feldspars *Radiat. Meas.* **23** 281–5

[22] Pagonis V, Kitis G and Furetta C 2006 *Numerical and Practical Exercises in Thermoluminescence* (Berlin: Springer Science)

[23] Dawam R R and Chithambo M L 2018 Thermoluminescence of annealed synthetic quartz: the influence of annealing on kinetic parameters and thermal quenching *Radiat. Meas.* **120** 47–52

[24] Agullo-Lopez F, Catlow C R A and Townsend P D 1988 *Point Defects in Materials* (London: Academic Press)

[25] Henderson B and Imbusch G F 2006 *Optical spectroscopy of inorganic solids* (Oxford: Oxford Science Publications) reprint

[26] Evans B D 1995 A review of the optical properties of anion lattice vacancies, and the electrical conduction in α-Al_2O_3: their relation to radiation-induced electrical degradation *J. Nucl. Mater.* **219** 202–23

[27] McKeever S W S, Moscovitch M and Townsend P D 1995 *Thermoluminescence dosimetry materials: Properties and uses* (Kent: Nuclear Technology Publishing)
[28] Yukihara E G and McKeever S W S 2011 *Optically Stimulated Luminescence: Fundamentals and Applications* (New York: Wiley)
[29] Chithambo M L 2004 Concerning secondary thermoluminescence peaks in a-Al_2O_3:C *South Afr. J. Sci.* **100** 1–4
[30] Chithambo M L, Seneza C and Ogundare F O 2014 Kinetic analysis of high temperature secondary thermoluminescence glow peaks in a-Al_2O_3:C *Radiat. Meas.* **66** 21–30
[31] Kortov V S, Milman I I, Moiseykin E V, Nikiforov S V and Ovchinnikov M M 2006 The possibility of using competing effects of deep traps in aluminium oxide for luminescent thermometry *Radiat. Prot. Dosim.* **199** 41
[32] Mishra D R, Kulkarni M S, Muthe K P, Thinaharan C, Roy M, Kulshreshtha S, Kannan K S, Bhatt B C, Gupta S K and Sharma D N 2007 Nanoparticles of Al_2O_3: Cr as a sensitive thermoluminescent material for high exposures of gamma rays irradiations *Radiat. Meas.* **42** 170
[33] Yukihara E G, Whitley V H, Polf J C, Klein D M, McKeever S W S, Akselrod A E and Akselrod M S 2003 The assessment of Al_2O_3:C responses as TL and OSL dosimeter in the dosimetery of charged particle of radiation fields *Radiat. Meas.* **37** 627–38
[34] Yazici A N, Solak S, Ozturk Z, Topaksu M and Yegingil Z 2003 The analysis of dosimetric thermoluminescent glow peak of α-Al_2O_3: C after different dose levels by β-irradiation *J. Phys. D: Appl. Phys.* **36** 181–91
[35] Chithambo M L, Seneza C and Kalita J M 2017 Phototransferred thermoluminescence of Al_2O_3:C: experimental results and empirical models *Radiat. Meas.* **105** 7–16
[36] Colyott L E, Akselrod M S and McKeever S W S 1996 Phototransferred thermoluminescence in alpha-Al_2O_3:C *Radiat. Prot. Dosim.* **65** 263–6
[37] Bulur E and Göksu H Y 1999 Phototransferred thermoluminescence from a-Al_2o_3:c using blue light emitting diodes *Radiat. Meas.* **30** 203
[38] Chithambo M L and Seneza C 2014 Kinetics and dosimetric features of secondary thermoluminescence in carbon-doped aluminium oxide *Physica B: Condens. Matter* **439** 165–8
[39] Polymeris G S, Raptis S, Afouxenidis D, Tsirliganis N C and Kitis G 2010 Thermally assisted OSL from deep traps in Al_2O_3:C *Radiat. Meas.* **45** 519–22
[40] Nyirenda A N, Chithambo M L and Polymeris G S 2016 On luminescence stimulated from deep traps using thermally-assisted time-resolved optical stimulation in a-Al_2O_3:C *Radiat. Meas.* **90** 109–12
[41] Alexander C S and McKeever S W S 1998 Phototransferred thermoluminescence *J. Phys. D: Appl. Phys.* **31** 2908–20
[42] Akselrod M S and Gorelova E A 1993 Deep traps in highly sensitive α-Al_2O_3:C TLD crystals *Nucl. Tracks Radiat. Meas.* **21** 143–6
[43] Chithambo M L, Nyirenda A N, Finch A A and Rawat N S 2015 Time-resolved optically stimulated luminescence and spectral emission features of α-Al_2O_3: C *Physica B: Condens. Matter* **437** 62–71
[44] Kitis G, Papadopoulos J G, Charalambous S and Tuyn J W 1994 The influence of heating rate on the response and trapping parameters of α-Al_2O_3:C *Radiat. Prot. Dosim.* **183** 183–90
[45] Zhu J, Muthe K P and Pandey R 2014 Stability and electronic properties of carbon in α-Al_2O_3 *J. Phys. Chem. Sol.* **75** 379–83
[46] McKeever S W S and Chen R 1997 Luminescence models *Radiat. Meas.* **27** 625–61

[47] Nyirenda A N 2013 Mechanisms of luminescence in α-Al_2O_3:C: Investigations using time-resolved optical stimulation and thermoluminescence techniques *Unpublished MSc thesis*

[48] Akselrod M S, Agersnap-Larsen N, Whitley V and McKeever S W S 1998 Thermal quenching of F-center luminescence in Al_2O_3:C *J. Appl. Phys.* **84** 3364–73

[49] Nikiforov S V, Milman I I and Kortov V S 2001 Thermal and optical ionization of F-centres in the luminescence mechanism of anion-defective corundum crystals *Radiat. Meas.* **33** 547–51

[50] Sashin V A, Bolorizadeh M A, Kheifets A S and Ford M J 2003 Electronic band structure of beryllium oxide *J. Phys. Condens. Matter* **15** 3567

[51] Ivanov V Y, Pustovarov V A, Kruzhalov A V and Shulgin B V 1989 Luminescence excitation of pure and impure BeO single crystals using synchrotron radiation *Nucl. Instrum. Meth. A.* **282** 559

[52] Tochlin E, Goldstein N and Miller W G 1969 Beryllium oxide as a thermoluminescent dosimeter *Health Phys.* **16**(1)

[53] Ogorodnikov I N and Kruzhalov A V 1997 Defect properties of beryllium oxide *Mater. Sci. Forum* **239-241** 51

[54] McKeehan L W 1922 The crystal structure of beryllium and of beryllium oxide *Proc. Natl Acad. Sci. USA* **8** 270

[55] Albrecht H O and Mandeville C E 1954 The phosphorescence of thoria *Phys. Rev.* **101** 1250

[56] Austerman S B 1963 The inversion twin: prototype in beryllium oxide *J. Nucl. Mater.* **14** 225

[57] Newkirk H W and Smith D K 1965 The (twin) composition plane as an extended defect and structure-building entity in crystals *Am. Miner.* **50** 22

[58] Mandeville C E and Albrecht H O 1954 Luminescence of beryllium oxide *Phys. Rev.* **94** 494

[59] Sommer M and Henniger J 2006 Investigation of a BeO-based optically stimulated luminescence dosemeter *Radiat. Prot. Dosim.* **119** 394

[60] Scarpa G 1970 The dosimetric use of beryllium oxide as a thermoluminescent material: a preliminary study *Phys. Med. Biol.* **15** 667

[61] Crase K W and Gammage R B 1975 Improvements in the use of ceramic BeO for TLD *Health Phys.* **29** 739

[62] Busuoli G, Lembo L, Nanni R and Sermenghi I 1983 Use of BeO in routine personnel dosimetry *Radiat. Prot. Dosim.* **6** 317

[63] Antonov-Romanovfsky V, Keirum-Markus I, Poroshina M and Trapeznikova Proc Z 1955 *Conf. Academy of Sciences of the USSR on the Peaceful Uses of Atomic Energy* **1956** p 239

[64] Hazen R M and Finger L W 1986 High-pressure and high-temperature crystal chemistry of beryllium oxide *J. Appl. Phys.* **59** 3728

[65] Petrovich J J Haertling and Carol Lynn Los Alamos National Laboratory Report *Beryllium Oxide (BeO) Handbook LA-UR-20-24561*

[66] Antsigin I N and Kruzhalov A V 1995 Lattice defects in beryllium oxide *Rad. Effects. Def. Sol.* **134** 303

[67] Crawford Jr J H and Slifkin L M (ed) 1972 *Point Defects in Solids* (Berlin: Springer)

[68] Song J, Liu T, Shi C, Sun R and Wu K 2021 *Mod. Phys. Lett. B.* **35** 2150148

[69] Yukihara E G 2020 A review on the OSL of BeO in light of recent discoveries: the missing piece of the puzzle? *Radiat. Meas.* **134** 106291

[70] Bulur E and Göksu H Y 1998 OSL from BeO ceramic: new observations from an old material *Radiat. Meas.* **29** 639

[71] Bulur E 2007 Photo-transferred luminescence from BeO ceramics *Radiat. Meas.* **42** 334

[72] Isik M, Bulur E and Gasanly N M 2017 Photo-transferred thermoluminescence of shallow traps in b-irradiated BeO ceramics *J. Lumin.* **187** 290

[73] Chithambo M L and Kalita J M 2021 Phototransferred thermoluminescence of BeO: time-response profiles and mechanisms *J. Appl. Phys.* **130** 195101

[74] Yukihara E G 2019 Observation of strong thermally transferred optically stimulated luminescence (TT-OSL) in BeO *Radiat. Meas.* **121** 103–8

[75] Yukihara E G 2011 Luminescence properties of BeO optically stimulated luminescence (OSL) detectors *Radiat. Meas.* **46** 580–7

[76] Yukihara E G 2019 Characterization of the thermally transferred optically stimulated luminescence (TT-OSL) of BeO *Radiat. Meas.* **126** 106132

[77] Ogorodnikov I N, Yu. Ivanov V and Kruzhalov A V 1995 Short-wavelenght luminescence and thermostimulated processes in single crystals of BeO *Radiat. Meas.* **24** 417

[78] Lontsi Sob A J, Chithambo M L and Kalita J M 2021 Analysis of illumination-time-dependent profiles of phototransferred thermoluminescence of Al_2O_3: C, Mg *J. Lumin.* **230** 117721

[79] Nikiforov S V, Gerasimov M F, Ananchenko D V, Shtang T V and Nikiforov A F 2022 Isothermal decay of thermoluminescence and energy distribution of traps in Al_2O_3–BeO ceramic *Radiat. Meas.* **153** 106752

[80] Avdyushin I G, Nikiforov S V, Kiryakov A N and Nikiforov A F 2018 *AIP Conf. Proc.* **2015** Article 020004

[81] Nikiforov S V, Avdyushin I G, Ananchenko D V, Kiryakov A N and Nikiforov A F 2018 Thermoluminescence of new Al_2O_3–BeO ceramics after exposure to high radiation doses *Radiat. Isot.* **141** 15

[82] Nikiforov S V, Borbolin A D, Marfin A Y, Ananchenko D V and Zvonarev S V 2020 New luminescent ceramics based on anion-deficient Al_2O_3–BeO for high-dose dosimetry *Radiat. Meas.* **134** 106303

[83] Santos A J J, de Lima J F and Valerio M E G 2001 Phototransferred thermoluminescence of quartz *Radiat. Meas.* **33** 427

[84] Shinsho K, Kawaji Y, Yanagisawa S, Otsubo K, Koba Y, Wakabayashi G, Matsumoto K and Ushiba H 2016 X-ray imaging using the thermoluminescent properties of commercial Al_2O_3 ceramic plates *Appl. Radiat. Isotope.* **111** 117

[85] Bhadane M S, Dahiwale S S, Sature K R, Patil B J, Bhoraskar V N and Dhole S D 2017 TL studies of a sensitive $CaNa_2(SO_4)_2$:Dy nanophoshor for gamma dosimetry *Radiat. Meas.* **96** 1

[86] Bilski P, Obryk B, Gieszczyk W and Baran P 2020 Position of LiF: Mg, Cu, P TL peak as an alternative method for ultra-high-dose dosimetry *Radiat. Meas.* **139** 106486

[87] Shinsho K, Kawachi Y, Yanakisawa S, Otsubo K, Koba Y, Wakabayashi G, Matsumoto K and Ushiba H 2016 X-ray imaging using the thermoluminescent properties of commercial Al_2O_3 ceramic plates *Appl. Radiat. Isot.* **111** 117

[88] Shinsho K, Maruyama D, Yanagisawa S, Koba Y, Kakuta M, Matsumoto K, Ushiba H and Andoh T 2018 Thermoluminescence properties for x-ray of Cr-doped Al_2O_3 ceramics *Sensors Mater.* **30** 1591

[89] Yanagisawa S, Maruyama D, Oh R, Koba Y, Andoh T and Shinsho K 2020 Two-dimensional thermoluminescence dosimetry using Al_2O_3: Cr ceramics for 4, 6, and 10 MV x-ray beams *Sensors Mater.* **32** 1479

[90] Shimomura R *et al* 2020 Thermoluminescence efficiency and glow curves of Cr-doped Al_2O_3 ceramic TLD for a wide linear energy transfer range *Radiat. Meas.* **134** 106356

[91] Oh R, Yanagisawa S, Tanaka H, Takata T, Wakabayashi G, Tanaka M, Sugioka N, Koba Y and Shinsho K 2021 Thermal neutron measurements using thermoluminescence phosphor Cr-doped Al_2O_3 and Cd neutron converter *Sensors Mater.* **33** 2129
[92] Koba Y, Shimomura R, Chang W, Shinsho K, Yanagisawa S, Wakabayashi G, Matsumoto K, Ushiba H and Ando T 2018 Dose linearity and linear energy transfer dependence of Cr-doped Al_2O_3 ceramic thermoluminescence detector *Sensors Mater.* **30** 1599
[93] Chithambo M L, Shinsho K and Polymeris G 2023 Properties of phototransferred thermoluminescence of Al_2O_3:Cr *Physica B: Condens. Matter* **650** 414576
[94] Chithambo M L, Kalita J M and Finch A 2020 F- and F^+-band radioluminescence and the influence of annealing on its emission spectra in Al_2O_3:C,Mg *Radiat. Meas.* **134** 106306
[95] Kalita J and Chithambo M L 2018 Thermoluminescence of α-Al_2O_3:C,Mg annealed at 1200 °C *Nucl. Instrum. Meth.* B **422** 78–84
[96] Akselrod M S, Akselrod A E, Orlov S S, Sanyal S and Underwood T H 2003 Fluorescent aluminum oxide crystals for volumetric optical data storage and imaging applications *J. Fluores.* **13** 503
[97] Sykora G J and Akselrod M S 2010 Photoluminescence study of photochromically and radiochromically transformed Al_2O_3: C,Mg crystals used for fluorescent nuclear track detectors *Radiat. Meas.* **45** 631
[98] Rodriguez M G, Denis G, Akselrod M S, Underwood T H and Yukihara E G 2011 Thermoluminescence, optically stimulated luminescence and radioluminescence properties of Al_2O_3: C,Mg *Radiat. Meas.* **46** 1469
[99] Trindade N M and Jacobsohn L G 2018 Thermoluminescence and radioluminescence of α-Al_2O_3: C,Mg at high temperatures *J. Lumin.* **204** 598
[100] Akselrod M S and Akselrod A E 2006 New Al_2O_3: C,Mg crystals for radiophotoluminescent dosimetry and optical imaging *Radiat. Prot. Dosim.* **119** 218
[101] Saharin N S, Wagiran H and Tamuri A R 2014 Thermoluminescence (TL) properties of Al_2O_3: C, Mg exposed to cobalt-60 gamma radiation doses *Radiat. Meas.* **70** 11
[102] Trindade N M, Magalhães M G, Nunes M C S, Yoshimura E M and Jacobsohn L G 2020 Thermoluminescence of UV-irradiated α-Al_2O_3: C,Mg *J. Lumin.* **223** 117195
[103] Munoz J M, Lima L S, Yoshimura E M, Jacobsohn L G and Trindade N M 2021 OSL response of α-Al_2O_3: C, Mg exposed to beta and UVC radiation: a comparative investigation *J. Lumin.* **236** 118058
[104] Kalita J M and Chithambo M L 2018 Thermoluminescence of α-Al_2O_3:C,Mg annealed at 1200 °C *Nucl. Instrum. Methods Phys. Res.* B **422** 78
[105] Kalita J M and Chithambo M L 2017 Comprehensive kinetic analysis of thermoluminescence peaks of α-Al_2O_3: C,Mg *Luminescence* **185** 72
[106] Chithambo M L, Sob J L and Kalita J M 2022 Light-induced inter-electron-trap charge movement in annealed Al_2O_3: C,Mg *Physica* B **624** 413438
[107] Kalita J M and Chithambo M L 2019 Phototransferred thermoluminescence and thermally-assisted optically stimulated luminescence dosimetry using α-Al_2O_3: C,Mg annealed at 1200 °C *J. Lumin.* **205** 1
[108] Kalita J and Chithambo M L 2018 The effect of annealing and beta irradiation on thermoluminescence spectra of α-Al_2O_3:C,Mg *J. Lumin.* **196** 195–200
[109] Chithambo M L, Kalita J M and Trindade N M 2022 Processes related to phototransfer under blue-and green-light illumination in annealed Al_2O_3: C, Mg *J. Appl. Phys.* **131** 245101

[110] Chandler P J and Townsend P D 1979 Implantation temperature measurement using impurity luminescence *Radiat. Eff. Defect. Solid.* **43** 61

[111] Chithambo M L, Raymond S G, Calderon T and Townsend P D 1995 Low temperature luminescence of transition metal-doped beryls *J. Afr. Earth Sci.* **20** 53–60

[112] Luff B J and Townsend P D 1993 High sensitivity thermoluminescence spectrometer *Meas. Sci. Technol.* **4** 65–71

[113] Finch A A, Wang Y, Townsend P D and Ingle M 2019 A high sensitivity system for luminescence measurement of materials luminescence *J. Bio. Chem. Lumin.* **34** 280–9

[114] Chithambo M L and Costin G 2017 Temperature-dependence of time-resolved optically stimulated luminescence and composition heterogeneity of synthetic α-Al_2O_3:C *J. Lumin.* **182** 252

[115] Kalita J M and Chithambo M L 2022 Concerning a hole trap in α-Al_2O_3:C,Mg *J. Appl. Phys.* **132** 015103

[116] Yukihara E G, Bos A J J, Bilski P and McKeever S W S 2022 The quest for new thermoluminescence and optically stimulated luminescence materials: needs, strategies and pitfalls *Radiat. Meas.* **158** 106846

IOP Publishing

Phototransferred Thermoluminescence

Makaiko L Chithambo

Chapter 5

Natural materials

The thermoluminescence characteristics of natural minerals are diverse. Their glow curves and emission spectra depend on a number of factors including the occurrence, chemical composition and provenance of the mineral. Laboratory treatment such as annealing changes the glow peak count, modifies the emission spectra and alters the luminescence sensitivity. We can expect the phototransferred thermoluminescence to be just as unpredictable and that is the appeal. This chapter presents the phototransferred thermoluminescence of selected natural materials. The topics covered include measurement, analysis and mechanisms of phototransfer.

5.1 Quartz

5.1.1 Introduction

Quartz is one of the most abundant minerals in the Earth's crust second only to feldspar [1]. Quartz has metaphorically been a popular test ground for its stimulated luminescence whose features are influenced by various factors including geological origin, thermal provenance, impurity type and content. The literature on conventional thermoluminescence (TL) of quartz is extensive. Indeed, as many studies and texts (e.g. [1–5]) attest, there has been no let up in reports of conventional TL of quartz, with the enduring draw being such issues as measurement techniques, sensitivity, analytical methods, kinetic analysis, development of models to explain the TL and luminescence-based applications. In comparison, the body of work devoted to phototransferred thermoluminescence (PTTL) of quartz is meagre. The publication of such work has not only been sporadic but many such studies have tended to be exploratory and qualitative.

One of the earliest attempts to utilize PTTL for applications goes back to Bailiff *et al* [6], who considered its use for retrospective dosimetry on the basis that the concentration of residual charge at a supposed donor electron trap in a material reflects its archaeological age. The unstated assumption here is that the signal monitored originates from a single donor. As a case in point, Bailiff *et al* [6]

doi:10.1088/978-0-7503-3831-8ch5

monitored PTTL at the '110 °C' peak in quartz and assumed that of the possible alternatives activating at 210, 325 and 375 °C, the active donor was the last but one. The possible role of this '325 °C' electron trap as a likely source of phototransferred electrons had been mooted long before [7], investigated [8–11] and, in some cases, accepted as fact [12]. Concerning dosimetry, Benny and Bhatt [13] studied the PTTL of quartz with glow peaks at 110 °C, 220 °C and 370 °C, all of which, except for the one at 220 °C, could be reproduced under phototransfer. This absence of phototransfer at one peak, although not discussed in their report, exemplifies competition effects, an often-overlooked aspect of PTTL studies.

Some studies on annealed natural quartz [13–15] and annealed synthetic quartz (e.g. [16]) also attribute any PTTL observed to electrons originating from the 325 °C peak. When Bertucci *et al* [16] observed PTTL in quartz heated up to 700 °C (in a furnace) they deduced that PTTL in quartz is contributed to by deep electron traps. The same conclusion was drawn by Morris and McKeever [17], Bailiff *et al* [6] and Chithambo *et al* [18–20]. In these examples, the dependence of the PTTL intensity on duration of illumination used to induce phototransfer is ascribed to the restorative effect of the phototransfer.

The change of intensity with illumination for PTTL measured from quartz was formulated by Wintle and Murray [11] in terms of an idealized system of one acceptor (the '110 °C' peak) and one donor (the '325 °C' electron trap) under first order kinetics. This arrangement discounts any role of the deep or any other electron trap. The same problem was addressed in greater detail by Alexander and Mckeever [21], who used numerical simulation to predict how the PTTL intensity would evolve with duration of illumination for a system of one donor and one acceptor. This study extended the methodology to include the effect of non-radiative recombination in the process. In this method, the transfer of electrons from the donor to the acceptor or any other electron traps during illumination or heating to produce PTTL is formulated as coupled non-linear differential equations because retrapping is taken into account. These equations cannot be solved analytically and require numerical approximation to a solution. By varying values of certain variables, for example trap occupancy, McKeever and co-workers [21, 22] were able to numerically obtain various types of time-response curves. It should be pointed out that resembling profiles can be mathematically generated for the same system for various combinations of rates of optical stimulation from electron traps. Thus, results produced by numerical simulation with one set of parameters can also be found when the permutation involves a different set of parameters (e.g. [23]). There are then three options to the validity of the sets of parameters—either both sets are valid, or one is, or indeed the exercise is academic. The equivalent of this concept from vector plots is selecting one of infinite trajectories, any of which represents a possible solution.

The notion that analysis reliant on assumption may not relate to experimental results is just as valid for analytical solutions. If one supposes, for a system of two electron traps, that the acceptor optically loses charge during illumination but that there is negligible backscattering from the conduction band during this step, the solution of the coupled rate equations for the process is an intensity profile that goes through a peak with illumination time. If the assumption does not hold, the intensity profile asymptotically saturates. The point here is that analytical solutions used to account for time-dependence of PTTL may or may not be valid for the systems at hand if the attendant assumptions are invalid. The analysis of PTTL must therefore be informed by experiment and any assumptions must be tested.

An empirical model for PTTL was described by Chithambo *et al* [24]. This method can be applied to a system with any number of donors. Phototransfer from donors to an acceptor is described by sets of coupled linear or non-linear differential equations at the illumination stage only. The resulting analytical solutions can be applied to experimental data. The number or identity of the donor is not assumed. Instead, information on the number and importance of donors, and whether retrapping has any significance, is based on experiment.

The PTTL of annealed natural quartz has been studied elsewhere on quartz annealed at 1000 °C [18] and at 500 °C [25]. These and other studies show that the behaviour of PTTL from quartz is not universal and its responses may be influenced by many factors as specified earlier. Indeed, annealing quartz alters its luminescence sensitivity [26], lifetimes [27–29] and emission bands [30]. In this part, we use the work of Chithambo *et al* [18] as a case to demonstrate one way to measure and analyse PTTL from natural quartz. We discuss PTTL induced by 470 nm blue light and detected in the 250–390 nm band.

5.1.2 Crystalline structure

Quartz is a phase of crystalline silica and exists in low temperature form as α-quartz and in high temperature modification as β-quartz [1]. The inversion from α-quartz to β-quartz occurs at 573 °C and pressure 105 Pa [31]. The crystal structure of α-quartz, the ubiquitous form in stimulated luminescence studies, consists of SiO_4 tetrahedra linked at their corners with other tetrahedra. The crystal occurs as a hexagonal prism terminated by rhombohedral faces. A polyhedral model displaying the features as explained is shown in figure 5.1. The structural form of quartz is discussed in greater detail elsewhere (e.g. [1, 31]).

Crystalline structures throughout this chapter are shown as polyhedra rather than as ball-and-stick models. A polyhedral model displays the large-scale perspective of a structure in order to dispense with the minutiae of chemical bond linkages between individual atoms. In this way, the model draws attention to the bonded framework. The precise location of individual atoms is irrelevant.

5.1.3 Glow curve

Figure 5.2 presents a glow curve obtained after beta irradiation to 300 Gy. The dose is intentionally high to induce intense enough PTTL. The need for such high signals creates a necessary trade-off between combinations of either short optical wavelength and low dose or longer wavelength and higher dose.

There are four nominal glow peaks in figure 5.2 labelled I through IV; two intense ones at 70 °C and 170 °C (I and III), and weaker intensity components near 125 °C (II) and 300 °C (IV). In keeping with the style of the preceding chapter, PTTL is

Figure 5.1. A polyhedral model of the crystal structure of α-quartz.

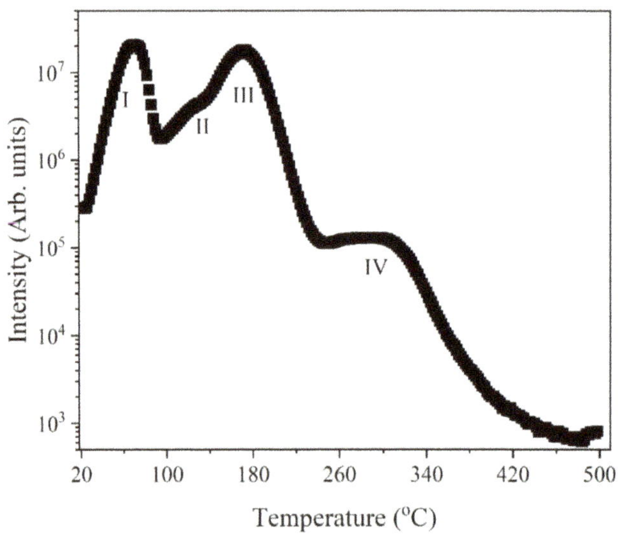

Figure 5.2. A glow curve measured at 1 °C s^{-1} after irradiation to 300 Gy [18]. Approximate positions of glow peaks are indicated. Reprinted from [18]. Copyright (2019) with permission from Elsevier.

described in relation to measurements made following preheating to remove each of the peaks in turn. This gives some indication on whether or not each peak not removed by preheating is associated with a donor electron trap. A glow curve recorded after the removal of peaks I and II is shown in figure 5.3. Only peak I is

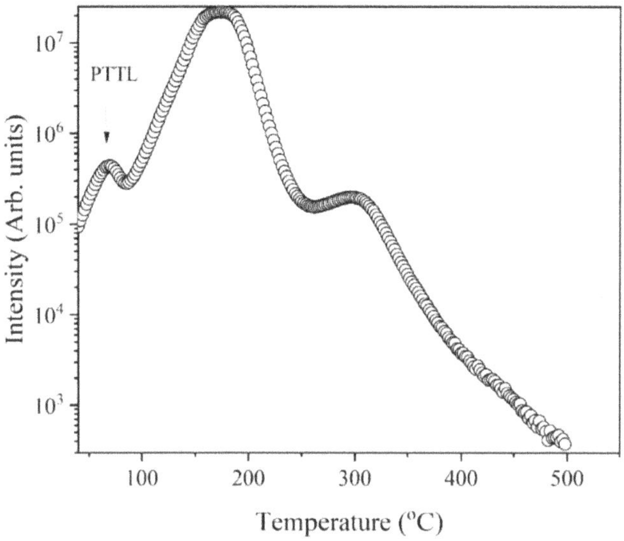

Figure 5.3. When peaks I and II are removed before illumination, phototransfer reproduces peak I.

reproduced under phototransfer as indicated. In general, the only peaks that reappear under phototransfer after various preheats are I, II and III. Peak II is not regenerated under phototransfer when all peaks in the glow curve are cleared off first.

5.1.4 Pulse annealing

Any peaks in a temperature-dependent set of signals, a glow curve, indicate the emptying of electron traps. The aim of a pulse annealing experiment is to go further and identify which ones are acting as acceptors or donors in the PTTL process. To do this, the PTTL corresponding to constant duration of illumination is monitored as the preheating temperature is progressively increased. Figure 5.4 shows such analysis for peaks I, III and IV. Peak I continually decreases with preheating temperature. One supposes that the decrease is caused by the depletion of its donors, namely, electron traps for peaks II–IV. In comparison, the change for peak III reflects both its role as a donor and the effect of preheating on its intensity. Specifically, between 80 °C and 140 °C as well as between 200 °C and 280 °C or so, the intensity is independent of preheating. Since the position of peak III is well above 140 °C, any preheating below this temperature has little effect on its intensity. The decrease between 140 °C and 200 °C is caused by the peak being partially removed by preheating.

Between 220 °C and 250 °C, the intensity of peak III remains constant because preheating hardly depletes its donor electron traps. One such donor corresponds to peak IV whose intensity is also independent of preheating up to 250 °C. When preheating depresses peak IV, that of peak III must also decline and it does.

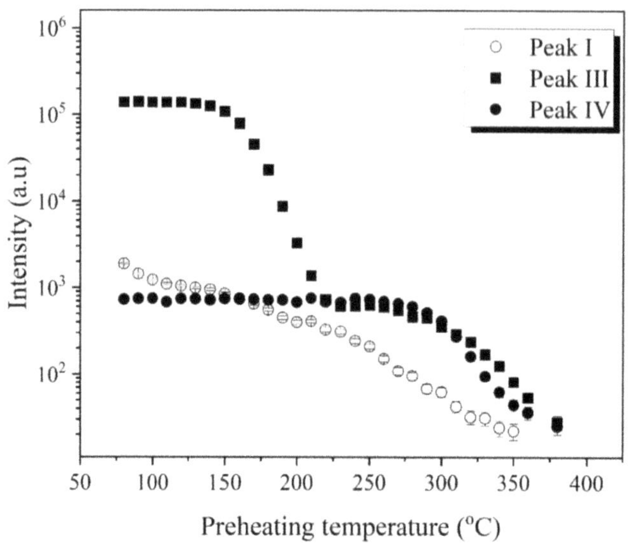

Figure 5.4. The influence of preheating on peak intensity in measurements made on a sample irradiated to 300 Gy, preheated in turn to temperatures between 80 °C and 380 °C in steps of 10 °C and illuminated for 60 s each time [18].

5.1.5 A qualitative assessment of PTTL time-response profiles

Preparatory to mathematical analysis of PTTL time-response curves, it is instructive to develop a qualitative understanding of the process. In particular, the assumption that any peak not removed by preheating corresponds to a donor needs to be experimentally verified. This is because the number of donors is a factor when setting up rate equations that describe the PTTL process. With reference to positions of peaks shown in figure 5.2, we discuss PTTL monitored whenever each of the peaks I through IV are removed in turn. Of particular interest is that preheating to 500 °C is a means to explore the role of deep electron traps in the process. By analysing the PTTL systematically this way, one can have a better idea of how to link various peaks under phototransfer.

5.1.5.1 PTTL following removal of peak I

When peak I is the only one removed by preheating prior to illumination, it is reproduced under phototransfer. Figure 5.5(a) shows that its intensity passes through a maximum with duration of illumination. The qualitative interpretation of this behaviour is that the intensity increases if more electrons are transferred to its electron trap than are removed by optical stimulation. When more electrons are lost than are retained, the intensity decreases.

5.1.5.2 PTTL following preheating that removes peaks I and II

When peaks I and II are removed, only peak I, not both, reappears under phototransfer. The time-dependence of its PTTL intensity resembles that of

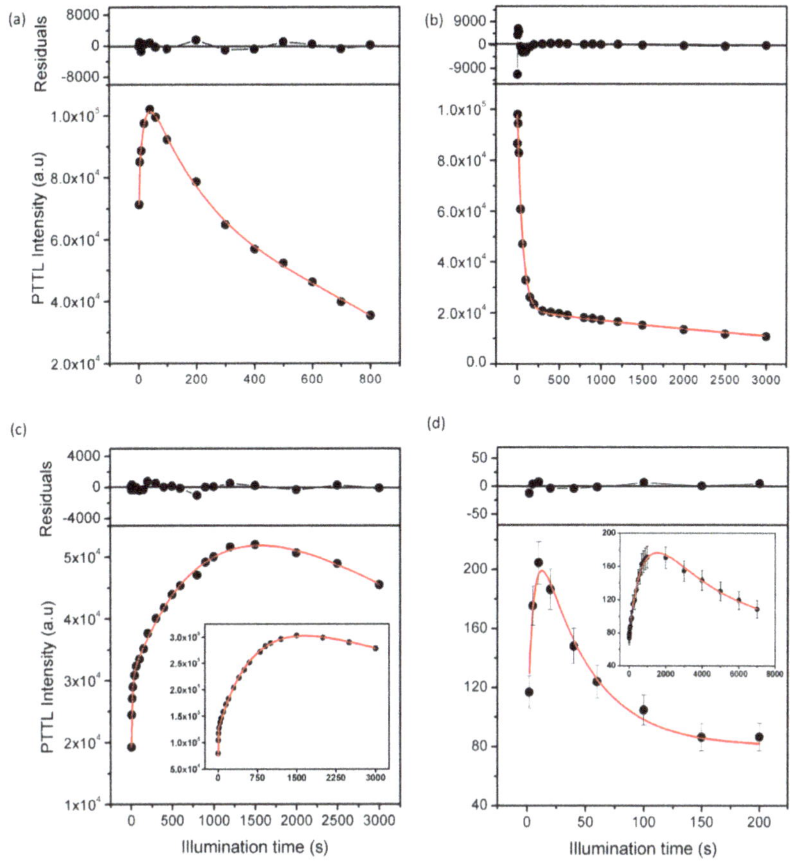

Figure 5.5. Contrasting characteristics of the dependence of PTTL intensity on duration of illumination. The measurements shown are for peak I (a, b and d), peak II (c) and peak III (insets to c and d). The residuals relate to the mathematical descriptions in the text. Reprinted from [18], Copyright (2019), with permission from Elsevier.

figure 5.5(a) except that its intensity is greater [18]. This is likely due to a portion of charge which would have otherwise photoinduced peak II. The absence of PTTL at peak II cautions that we need not always view PTTL as a restorative effect.

5.1.5.3 PTTL following preheating that removes peaks I–III

After preheating to remove peaks I–III, all three are reproduced under phototransfer. Figure 5.5(b) shows the change for peak I whose intensity decreases consistently with illumination. Chithambo *et al* [18] ascribed this behaviour to competitive retrapping at electron traps for peaks II and III. They reasoned that if retrapping causes fewer electrons to be captured at the acceptor each time the illumination is lengthened, the intensity of the resulting PTTL will consistently decrease.

The patterns for peaks II and III in figure 5.5(c) where the changes are drawn out can also be thought of in terms of retrapping. This is discussed to be an effect of their electron traps being effective charge competitors [18].

5.1.5.4 PTTL from deep electron traps

One way to monitor PTTL due to deep electron traps is to first raise the preheating temperature to 500 °C or more before illumination. The choice of 500 °C is arbitrary and so is the definition of a deep trap as any that activates beyond this limit. The outcome of such a protocol is shown in figure 5.5(d). In this example, PTTL only appears at peaks I and III.

5.1.6 Mathematical analysis

PTTL can be analysed on the basis of theoretical conceptions by invoking suitable assumptions or wholly by empiricism. The assumptions must be experimentally verified. We will use the preceding qualitative descriptions which can be used as a basis for an empirical model. PTTL is measured in several steps starting with irradiation to populate electron traps, preheating to deplete some, illumination to transfer electrons from intact deeper traps to emptied shallower ones, and heating to monitor PTTL. Thus, the number of electron traps that act as acceptors or donors (source traps) changes with preheating temperature. However, it is inadvisable to assume that the electron traps of all peaks not affected by preheating are donors. Rate equations describing charge transfer between electron traps can be formulated on the basis that only a certain portion of electrons optically stimulated from a donor end up at an electron trap and that the transitions proceed via the conduction band. The set of rate equations describing inter-electron-trap transfer are set up at the illumination stage only.

Figure 5.6 shows an energy band diagram used to explain PTTL in this discussion. Electron traps for peaks I, II, III and IV are labelled as such. A deep electron trap, associated with PTTL that ensues after preheating to and beyond 500 °C, is

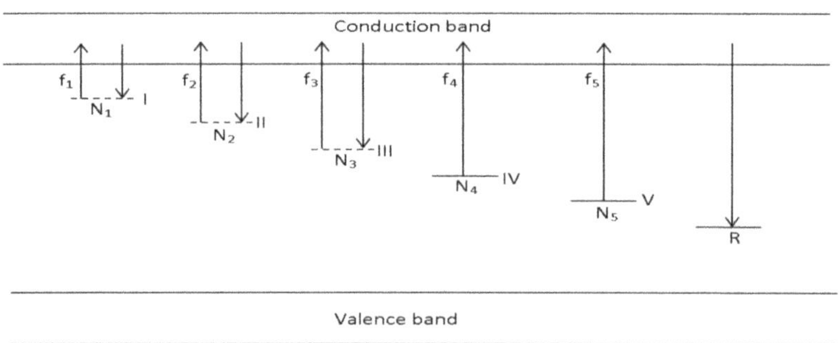

Figure 5.6. An energy band scheme used to describe PTTL. Electron traps are shown as I–V. The recombination centre R is included for completeness. The optical stimulation rates from electron traps are indicated as f_i and the concentration at each electron trap as $N_i(i = 1, \ldots, 5)$. Reprinted from [18], Copyright (2019), with permission from Elsevier.

indicated as V. Since the number of deep traps is unknown, the label V is representative. The rate of optical stimulation from a donor electron trap during illumination is expressed as $f = \Phi\sigma$ where Φ is the incident photon flux and σ the photoionization cross-section. Retrapping is neglected since I and III are first order peaks [18]. The resulting rate equations are linear and the set of coupled equations for a given acceptor have an analytical solution which can then be applied to experimental data.

The optically-induced movement of electrons from n donors to an acceptor can be expressed in matrix form as given in equation (3.17). The particular solution for the time-dependence of the PTTL for a system of one acceptor and n donors is given in equation (3.18). Figure 5.5(a) shows a fit of the model for peak I corresponding to preheating that removes this peak only. Since pulse annealing experiments show that level IV is redundant as a donor for PTTL at peak I, there are only three donors to consider and hence the PTTL can be discussed in terms of a system of three donors. The fit of equation (3.18) with $j = 3$ is shown in figure 5.5(a). Figure 5.5(b) shows the case for peak I following the removal of peaks I–III. Although nominally this is a system of two donors, the stimulation probability is $f_1 \gg f_4, f_5$. This reduces equation (5.2) to a sum of two simple exponential functions, namely

$$N_1 = C_4^* e^{-f_4 t} + C_5^* e^{-f_5 t} \tag{5.1}$$

The approximation describes the experimental result of figure 5.5(b). Figure 5.5(c) is for peaks II and III. Here, the PTTL corresponds to a system of one acceptor and two donors and the fits are each for $j = 2$. The final plots cases of peak I and III in figure 5.5(d) for peaks I and III are each for a system of one donor for which

$$N_k = C(e^{-f_1 t} - e^{-f_k t}) \tag{5.2}$$

The analysis described also offers a means to determine the photoionization cross-section. This is a measure of the ease with which an electron trap can be optically stimulated. The smaller this value, the less stable the electron trap. For example, for peak I, whose system has four donors, one of which makes a negligible contribution, the photoionization cross-section is calculated as 6.40×10^{-18} cm^2 whereas for the same peak but for a system of one donor, the value is 1.10×10^{-18} cm^2 [18].

5.1.7 Competition effects

When electrons are optically released from donors, they can move to recombination centres, backscatter or transit to acceptors. Although this idealized assumption facilitates mathematical analysis, phototransfer can also be affected by competition effects. These are processes that enhance or depress the trapping of electrons during illumination.

In the quartz of figure 5.2, competition effects are observed after preheating that removes all peaks but better so for preheating, which clears peaks I and II only. Although competition effects can be discussed with reference to either of these two choices, we do so with respect to peak III. When the quartz is preheated to remove peaks I and II, the electron trap for peak III should be a donor for any PTTL.

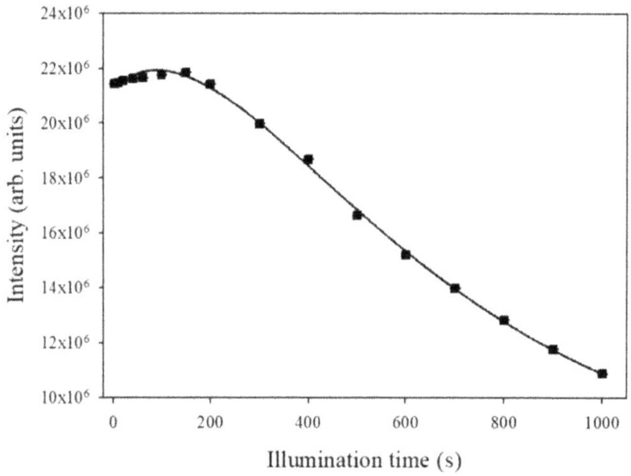

Figure 5.7. The change of PTTL intensity with illumination for peak III obtained where the sample is preheated to remove peaks I and II before illumination. The intensity does not decrease monotonically as might be expected for a donor peak but uncharacteristically goes up first. Reprinted from [18], Copyright (2019), with permission from Elsevier [18].

The dependence of the intensity of the conventional TL of peak III on duration of illumination is displayed in figure 5.7. The intensity does not decrease continually with illumination as might be expected of a donor but initially increases before the expected drop. Thus, the electron trap of peak III acts both as a competitor for phototransferred electrons from deeper electron traps and as a donor for PTTL at shallower electron traps.

Since the additional donors for the luminescence of peak III must be the electron traps IV and V (see figure 5.6), it is instructive to study the change that illumination causes in the intensity of peaks IV, the only one for which this can be done. The intensity decreases as expected of a donor [18]. The decrease is, however, slower than expected, which suggests that this electron trap also serves as an acceptor for electrons phototransferred from deep electron traps. Therefore, the processes causing the change at peak III (figure 5.7) including competition effects can be summarised as

$$\frac{dN_5}{dt} = -f_5 N_5 \tag{5.3}$$

$$\frac{dN_4}{dt} = -f_4 N_4 + \gamma_5 f_5 N_5 \tag{5.4}$$

$$\frac{dN_3}{dt} = -f_3 N_3 + \delta_4 f_4 N_4 + \delta_5 f_5 N_5 \tag{5.5}$$

Equation (5.3) expresses the optical stimulation of electrons from the deep electron trap. Equation (5.4) accounts for the optical detrapping of electrons from the electron trap of peak IV as well as competitive capture of some electrons from the deep trap. The first term in equation (5.5) is the loss of electrons from the acceptor (level III) during illumination. The second and third terms describe the trapping at level III of some of the electrons removed from levels IV and V. The solution of the family (5.3)–(5.5), giving the evolution with time of the intensity of peak III, is

$$N_3 = A_1(e^{-f_5 t} - e^{-f_3 t}) - A_2(e^{-f_4 t} - e^{-f_3 t}) + A_3(e^{-f_5 t} - e^{-f_3 t}) \tag{5.6}$$

where $A_1 = \delta_4 f_4 \gamma_5 N_{5i}/(f_4 - f_5)$, $A_2 = \delta_4 f_4 \gamma_5 N_{5i}/(f_3 - f_4)$, and $A_3 = \delta_5 f_5 N_{5i}/(f_3 - f_5)$. A fit of equation (5.6) is shown in figure 5.7.

It is also informative to consider the case of peak I, the only one reproduced under phototransfer when the first two peaks are preheated off. The system in question consists of level I as the acceptor and levels IV and V as donors. To describe the phototransfer adequately, it is important to address the question of whether competition effects at the donors are relevant. If the competition effects are included in the model, the expression for the PTTL time-response for peak I can be written as

$$N_1 = e^{-f_1 t} \int \sum_{j=3}^{5} \left(\Gamma_j s_j N_j\right) \; e^{f_1 t} dt + c\, e^{-f_1 t} \tag{5.7}$$

where Γ_j is the constant of proportionality, f_j is the probability of stimulation, N_j the occupancy at the jth electron trap and c is an integration constant. The particular solution can be simplified with recourse to an *ansatz* that competition effects are relevant only for peak III. If we treat levels III and IV as ordinary donors, we have with $j = 2$ in equation (3.18) a fit of which is shown in figure 5.8.

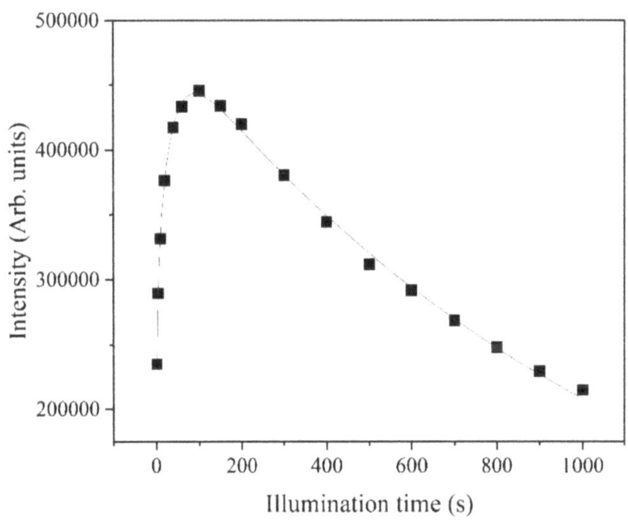

Figure 5.8. The time-dependence of PTTL for peak I after removal of peaks I and II by preheating. The line through the data is a fit corresponding to two donors. This is figure 9(c) in Ref. [18].

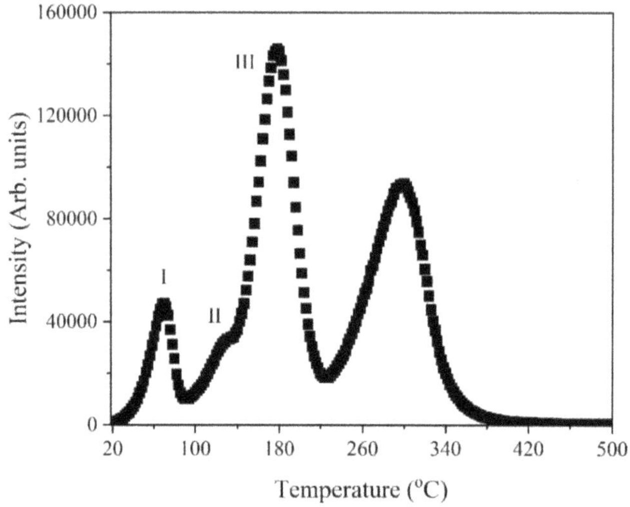

Figure 5.9. A glow curve recorded after preheating to remove peaks I–III followed by 470 nm illumination [18].

Figure 5.9 is a glow curve obtained after preheating that removes peaks I–III. All three are reproduced under phototransfer as labelled. Of these the third, not the first, is prominent unlike in figure 5.2 where the opposite is true. This result is consistent with figure 5.7, which shows that the electron trap for peak III is an effective competitor for phototransferred electrons.

5.1.8 Summary

The phototransfered thermoluminescence of quartz shows contrasting behaviour. While some peaks appear under phototransfer anyhow others only do so when certain other peaks have been removed. The emission can be linked to donors whose importance and number can be experimentally decided. Indeed, some supposed donors do not always act as such. An interesting feature of PTTL from quartz concerns competition effects, which affect the way the PTTL intensity scales with duration of illumination.

5.2 Tanzanite

5.2.1 Introduction

Tanzanite is a rare gem mineral found only at Merelani, Tanzania. Tanzanite occurs because of the confluence of unique geological features [32, 33]. The blue hue of tanzanite, which becomes more resplendent with heat treatment, and its trichroism lends it its commercially important aesthetic appeal. Tanzanite is mineralogically a zoisite. Zoisite, with the general formula $Ca_2Al_3(Si_2O_7)(SiO_4)O(OH)$, is an orthorhombic polymorph of clinozoisite [34]. The crystal structure of zoisite is shown in figure 5.10. The structure consists of rows of neighbouring octahedra that are

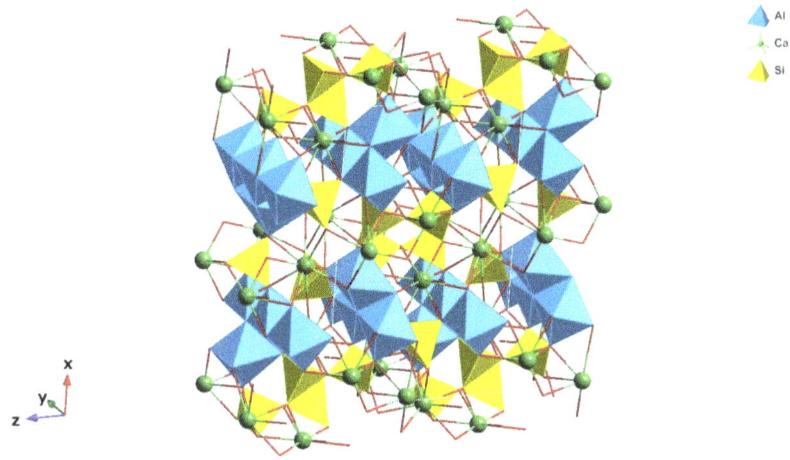

Figure 5.10. A polyhedral model of the crystal structure of zoisite. The crystal system is orthorhombic, space group *Pnma*. **There are two legends in this figure**. The one on the top right identifies Al, Ca and Si. The legend on the bottom left gives the orientation of the structure as drawn.

aligned parallel to the *y*-axis and linked by single SiO_4 and double Si_2O_7 tetrahedra groups. Sites occupied by Al have an octahedral coordination whereas Ca ions coordinated in a 9- or 10-fold way by oxygen occupy cavities within the structure. Although tanzanite is a sorosilicate, the substitution of Si by Al in the SiO_4 group occurs but only to a minor extent [35]. We can therefore expect that any stimulated luminescence processes in tanzanite will differ from those in silicates such as quartz where such replacement is important.

The TL of tanzanite has been studied elsewhere [36] where the dosimetric characteristics, kinetic analysis and mechanisms of TL in a beta-irradiated sample were reported. A glow curve measured at 1 °C s^{-1} after irradiation to 51 Gy showed nominal glow peaks near 70, 130 and 180 °C and an indistinct one at 240 °C. All these had features of first order kinetics. The influence of delay between irradiation and measurement on the glow peaks turned up notable findings. Peak I fades with delay between irradiation and measurement. In comparison, the intensities of peaks II and III are unaffected by such delay. When the irradiation dose is increased to 100 Gy, peaks II and III counterintuitively become brighter with delay between irradiation and measurement. When it comes to light-induced fading, all peaks fade when the tanzanite is illuminated by 470 nm blue light after irradiation [36]. Chithambo and Folley [36] attributed the appearance of peak I to electron recombination at $[TiO_4\backslash h^+]^\circ$ sites and optically stimulated luminescence to hole recombination at $[TiO_4]^-$ anions. The thermoluminescence of peaks II and III was reasoned to arise due to recombination of alkali cations at $[TiO_4]^-$ point defects [36].

By demonstrating that TL peaks in tanzanite are light sensitive and using the fact that the defect pair responsible for peak I is also produced during illumination, it was hypothesised [36] that tanzanite must produce phototransferred TL. That hypothesis was confirmed in preliminary tests in the same study [36] and

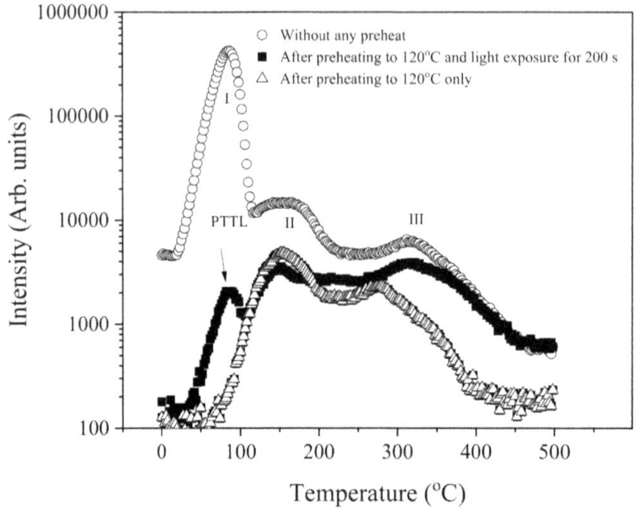

Figure 5.11. A glow curve measured at 1 °C s^{-1} immediately after irradiation (open symbols) compared with one recorded after irradiation and preheating to 120 °C to remove peak I and illumination. The latter shows peak I reproduced under phototransfer. No such peak is present in a glow curve recorded after preheating but without any illumination. Reprinted from [37], Copyright (2021), with permission from Elsevier.

substantively investigated subsequently [37]. That work, the only one on the PTTL of tanzanite, is the basis for the discussion in this section. The PTTL refers to phototransfer by 470 nm blue light and detection between 250 nm and 390 nm.

5.2.2 Glow curve

A glow curve obtained after irradiation to 120 Gy is shown in figure 5.11 (open circles). There are three glow peaks at 86 °C (I), 160 °C (II) and 320 °C (III). For comparison, a second glow curve (solid symbols) measured after the same irradiation, preheating to remove peak I, and illumination is included. By comparing this glow curve with a third one measured with similar settings except for illumination, it is easy to infer that peak I is reproduced by phototransfer

Of the four nominal peaks of tanzanite, peaks II and III in figure 5.11 may each be collocated glow peaks. These are peaks whose components are so closely spaced and embedded within one another to such an extent that the product appears single. Their components cannot be readily separated by thermal cleaning or by the T_m–T_{stop} partial heating technique [36]. Glow peaks of zoisite are known to share this property [38]. Tanzanite therefore exemplifies a case where the PTTL can be described as a system of one acceptor and an indeterminate number of donors. This is the motivation for studying the PTTL of tanzanite.

5.2.3 Identification of electron traps as donors and acceptors

To identify the role of electron traps as acceptors or donors, one uses pulse annealing as described earlier. The resulting influence of preheating on the intensity of peaks I,

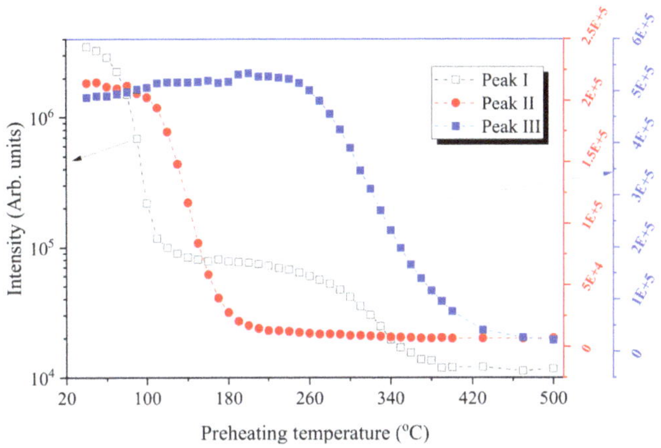

Figure 5.12. The influence of preheating temperature on the intensity of peaks I, II and III for measurements made between 40 °C and 500 °C at intervals of 10 °C and illumination for 100 s each time. Reprinted from [37], Copyright (2021), with permission from Elsevier.

II and III is displayed in figure 5.12. The patterns for peaks I and II are comparable. For both, the intensity is either independent of preheating or decreases monotonically with preheating. The decrease occurs when preheating gradually removes the peak itself or depletes the donor for its PTTL. In contrast, peak III increases in intensity with preheating up to about 200 °C, beyond which it decreases. This suggests that peak III acts as a competitor for electrons released from other electron traps over these preheating temperatures.

Figure 5.12 suggests that after preheating that removes only peak I, the main donor is the electron trap of peak II. On the other hand, when peaks I and II are cleared, the main donor for the PTTL is the electron trap for peak III. No PTTL appears when the tanzanite is preheated to 500 °C to remove all peaks.

5.2.4 Fading

The PTTL decreases with delay between illumination and measurement whether only peak I is heated off or whether both peaks I and II are (figure 5.13(a)). Examples of glow curves obtained after various delays (figure 5.13(b)) confirm the earlier observation that peaks II and III are not genuinely single peaks. The peaks shown in this figure are resolved into several components by deconvolution.

5.2.5 Time-response profiles

We now consider the time-response profiles for PTTL obtained after preheating to remove either peak I only or both peaks I and II. PTTL is obtained only in this way.

Figure 5.14 displays the effect of duration of illumination on the intensity of peak I obtained under phototransfer. The changes for the putative donor peaks II and III are descriptive. All three decrease in intensity with illumination. Peak II does so

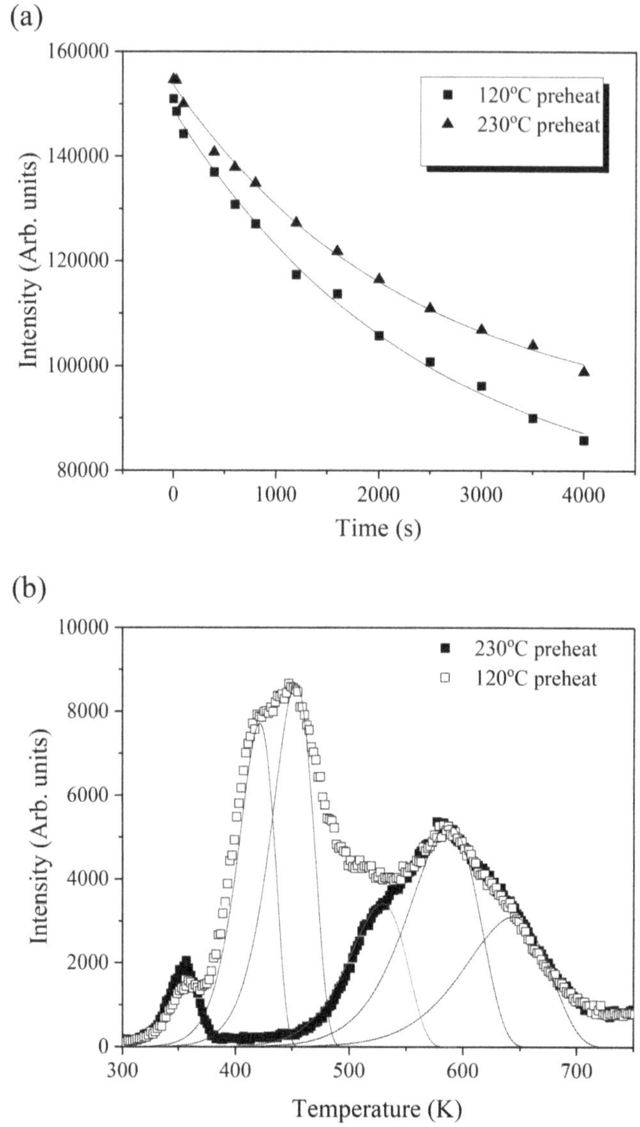

Figure 5.13. The influence of delay between irradiation and measurement on PTTL peak I reproduced after preheating to 120 °C or 230 °C (to remove either peak I only or peaks I–II) (a). Glow curves obtained after preheating to 230 °C and delay for 1600 s between irradiation and measurement (solid symbols) and another measured after preheating to 120 °C and delay for 3000 s are shown resolved into various components using deconvolution (b). This and the experimental form of the glow peaks show that the peaks are not *bona-fide* single ones. Reprinted from [37], Copyright (2021), with permission from Elsevier.

faster than peak III. Since the only donor that can be linked to PTTL after preheating to 120 °C (to remove peak I only) is the electron trap for peak II (figure 5.12), peak I weakens when illumination depletes the electron trap for peak II.

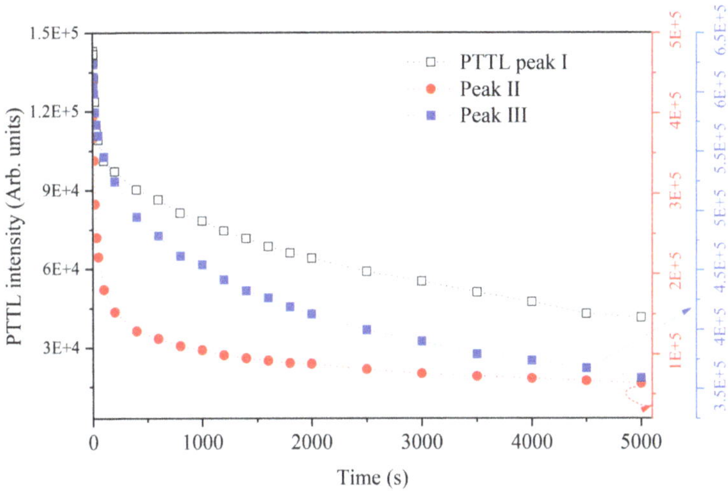

Figure 5.14. The PTTL time-response profile for peak I after preheating that removes this peak only. Results for peaks II and III are shown for comparison. Reprinted from [37], Copyright (2021), with permission from Elsevier.

5.2.6 Mathematical analysis of time-response profiles

The main question we wish to address here is how to approach the analysis of PTTL of collocated glow peaks as exemplified by tanzanite. We see two possibilities to this. Firstly, the PTTL can be described in terms of a system of one acceptor and two nominal donors whose contribution depends on the preheating temperature. One assumes that each collocated peak has a dominant component. The second way is to remain consistent with experiment and treat the PTTL as corresponding to a system of one acceptor and an indeterminate number of donors.

PTTL is only regenerated at peak I when the first two or all three peaks are cleared off prior to illumination. In either case, there is only one nominal donor for the PTTL. This is either the electron trap for peak II or that of peak III. With each choice, the PTTL can be described by a system of one acceptor and one donor. The typical result for such a system is an intensity profile that goes through a maximum and decreases to some stable value [11, 18–21, 23]. Notably, peak I in tanzanite shows a different response for the same system. Its intensity does not stabilize at long illumination times but rather continues to decrease consistently [37]. The discrepancy with the archetype may relate to the glow peaks in tanzanite being collocated composites. We reach the same conclusion when we consider the PTTL at peak I where with $j = 2$, equation (3.18) also improperly compares to experimental data. The PTTL of tanzanite should therefore be analysed in terms of a system of one acceptor and an indefinite number of donors. Such analysis is, however, yet to be developed. An interim empirical approach is to expand the sum in equation (3.18)

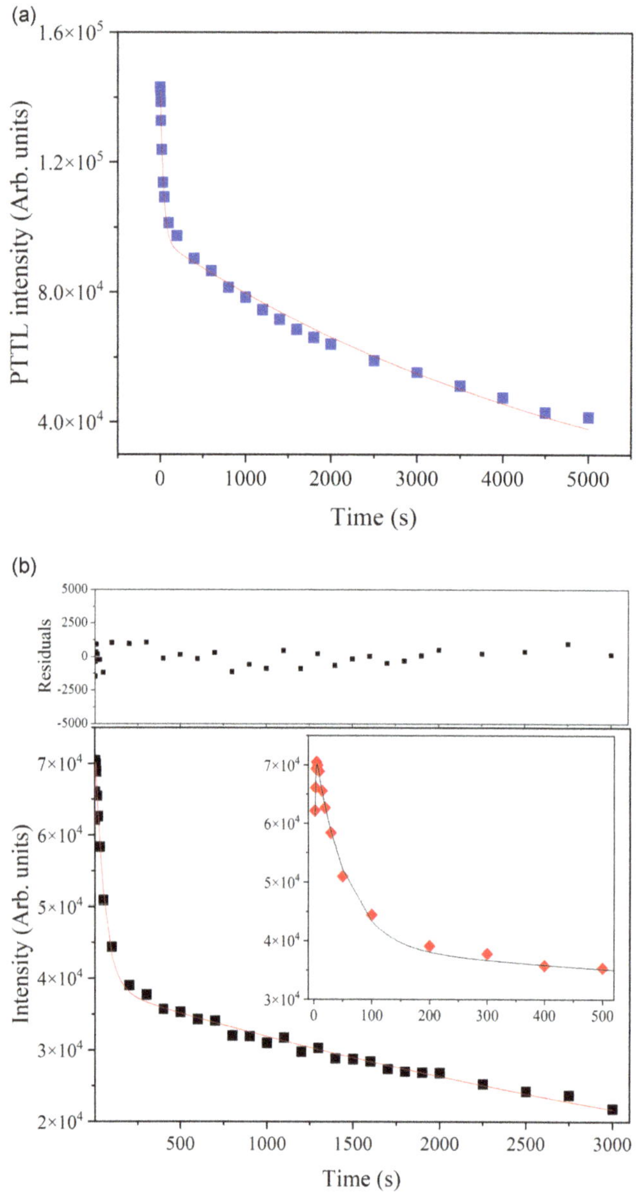

Figure 5.15. The intensity of PTTL intensity as a function of illumination time for peak I following preheating to 120 °°C (a) to 230 °°C (b). The inset in part (b) displays the change in the first 500 s. The lines through the data are mathematical descriptions given in the text. Reprinted from [37], Copyright (2021), with permission from Elsevier.

one at a time until a good fit is obtained. As an example, figure 5.15(a) displays the time-dependence of PTTL corresponding to preheating that removes peaks I and II fitted by equation (3.18) with $j = 2$ instead. Retrapping is neglected since nominal peaks I, II and III follow first order kinetics [36, 37].

5.2.7 Competition effects

As observed for quartz in the preceding section, competition effects can affect PTTL. These are experimental processes that tend to enhance or decrease the amount of phototransferred thermoluminescence. An instance of this is when the intensity of a donor counterintuitively increases. To discuss this for tanzanite, we single out peak III.

5.2.7.1 Qualitative account of competition effects

Phototransfered TL at peak I only appears when either or both peaks II and III are removed first. The associated pulse annealing result is given in figure 5.12. Here, the intensity of peak III, the supposed donor, unexpectedly becomes more intense with preheating in the first 200 °C or so.

If a glow peak appears beyond a certain temperature, it may be linked as a donor to any PTTL that appears below that temperature. If the contribution is weak, the intensity of that donor peak should either be independent of preheating or decrease with it. The change for peak II in figure 5.12 is in keeping with this conclusion. If, however, electrons competitively backscatter to a donor electron trap, the extent to which this occurs can be seen in a plot of intensity against preheating temperature. This is exemplified for peak III in the same figure 5.12 where its intensity initially increases with preheating temperature. Since the position of this peak is beyond any of the preheating temperatures, the increase in intensity must be a consequence of its electron trap acting as an acceptor in competition with the electron trap for peak I. It is only when the preheating temperatures go beyond 200 °C that the intensity decreases because, in that case, there is significant loss of trapped charge owing to the combined effect of preheating and illumination.

5.2.7.2 Empirical model of competition effects

We now consider figure 5.16 where the intensity of peak III is re-plotted over different scales to amplify certain parts of its temperature dependence. The x-axis is a dummy variable t' standing for the measurement number, say some integer n', where the initial count ($n' = 1$) corresponds to the first preheating temperature. These temperatures are listed on the over-absicca.

Figure 5.16 shows that the concentration of electrons at the electron trap of peak III is affected by preheating and illumination. The intensity of peak III increases in the first nine steps of preheating, levels off and then decreases consistently thereafter.

This behaviour can be accounted for by invoking the same arguments used to describe competition effects in synthetic quartz [19]. We relate the concentration of electrons N_3 at the electron trap for peak III to two independent parameters, a variable one N and a time-independent one N_A, that is, $N_3 = N + N_A$. The parameter N is the number of electrons optically stimulated to the conduction band from the donor (electron trap for peak II) and N_A is the number of electrons at the electron trap for peak III immediately after irradiation but before illumination. N_A is constant since it corresponds to identical irradiation. We also assume that electrons will not

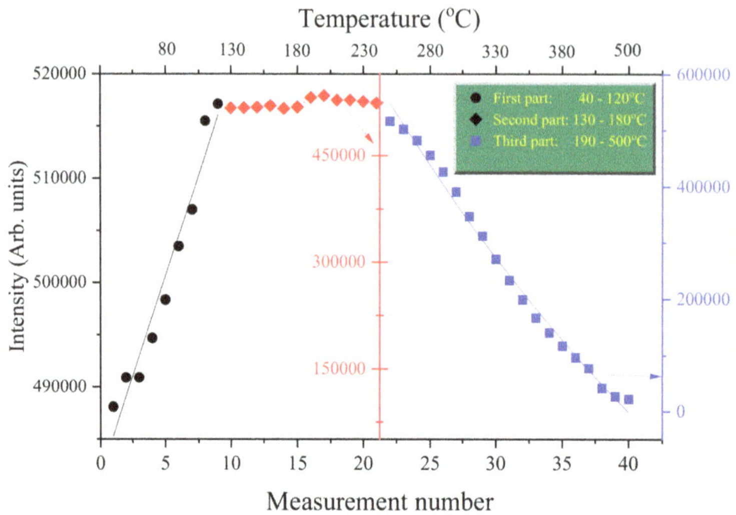

Figure 5.16. The change of intensity of peak III related to the measurement number. The temperatures corresponding to each counter on the bottom axis are shown on the over-axis. Reprinted from [37], Copyright (2021), with permission from Elsevier.

accumulate in the conduction band. These approximations, which only apply for temperatures at which peak II is not affected by preheating [37], are helpful for further analysis as now follows.

The concentration of electrons in the conduction band changes owing to illumination or irradiation. Since N_A remains constant and the sample undergoes identical illumination each time, N decreases because of competitive retrapping and simultaneous transfer to recombination centres.

5.2.7.3 Increase of signal with preheating

The concentration of electrons N in the conduction band following optical stimulation from the donor can be written as

$$N' = f_2 N_{2i} - \lambda N \tag{5.8}$$

where N_{2i} is the initial concentration of electrons at the donor and λ is the probability per unit time that a stimulated electron will transit the competitive route. Hence, the time-dependence of N is

$$N = \frac{(f_2 N_{2i})}{\lambda}[1 - e^{-\lambda t}] \tag{5.9}$$

Figure 5.16 (solid circles) shows that the intensity of peak III initially increases with preheating. Thus, λ is negligible and (5.9) simplifies to

$$N \approx (f_2 N_{2i})t \tag{5.10}$$

By replacing t in equation (5.11) by t' denoting the measurement number, we can write

$$N \approx (\alpha_2 f_2 N_{2i}) t' \tag{5.11}$$

where α_2 is a constant of proportionality. The term within brackets in equation (5.11) expresses the fact that whenever the sample is illuminated after preheating, only a portion of the ensuing electrons from the donor moves to the electron trap for peak III. If the term λN in equation (5.8) decreases with preheating (that is, with each measurement), the term within brackets in equation (5.11) will increase and, correspondingly, so will N with t' (i.e. N will increase with preheating). The line though the data between 40 °C and 120 °C (solid circles) in figure 5.16 is therefore a fit of equation (5.11).

5.2.7.4 Decrease of signal with preheating

If the sample is taken beyond 240 °C, the electron trap for peak III is depleted. The peak then becomes fainter with both preheating and illumination. These two processes are separate since the sample is preheated first and illuminated second. The dependence of charge loss on preheating cannot be modelled precisely. However, a qualitative account for the reduction of charge due to illumination can be made.

The change in the electron concentration at the electron trap for peak III can be expressed as

$$N_3' = -f_3 N_3 - \gamma_3 N_3 \tag{5.12}$$

where $f_3 N_3$ is the charge lost owing to optical stimulation and the term $\gamma_3 N_3$, where γ_3 is a constant of proportionality, expresses the effect of preheating. Assuming that the most change occurs owing to preheating so that $\gamma_3 \gg f_3$, it follows that

$$N_3 = N_{3i} e_{-\gamma_3 t'} (1 - f_3 t') \tag{5.13}$$

where all symbols are as defined earlier. A fit of equation (5.13) to experimental results is shown in figure 5.16.

5.2.7.5 Vector fields model of competition effects

The analytical solutions of the previous section rely on approximations because the way preheating reduces the concentration at electron traps cannot be modelled accurately. That approach can be complemented by using vector field analogues of equations (5.8) and (5.12) in order to display the general behaviour of the set of their solutions. This is done in figures 5.17(a) and (b), respectively. The inclination of the vector fields and any trajectories of possible solutions in the phase plane are consistent with the experimental results shown in the first and third segments of figure 5.16.

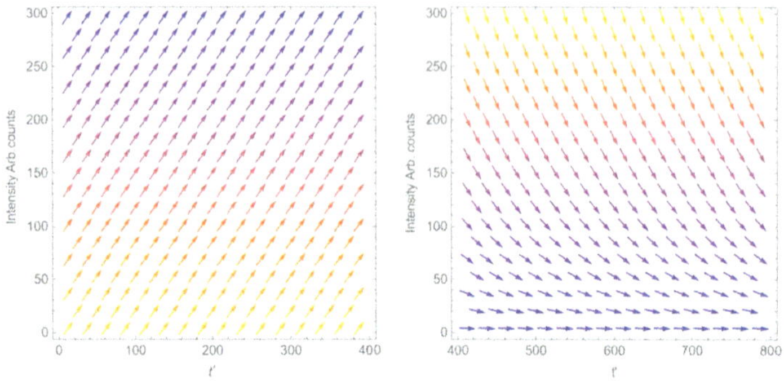

Figure 5.17. Vector fields of analogues of equations (5.8) and (5.12) intended to show the trajectories of their solutions. Reprinted from [37], Copyright (2021), with permission from Elsevier.

5.2.8 Summary

Phototransferred thermoluminescence can be induced by 470 nm illumination in the rare gem mineral, tanzanite. For heating at 1 °C s^{-1} after irradiation, its glow curve shows at least three clear peaks, a prominent one just below 90 °C (peak I) and two other weaker intensity ones near 160 °C and 320 °C. Only peak I is reproduced by phototransfer. The phototransferred thermoluminescence can be ascribed to an indeterminate number of donors. In practice, the dependence of the PTTL intensity on duration of illumination can be accounted for with a system of at least two donors. The PTTL is also affected by competition effects, where supposed donor electron traps act otherwise. Analysis based on empirical models is one instructive way to start to understand these processes.

5.3 CaF_2

5.3.1 Introduction

Fluorite, the mineral form of CaF_2, is a ubiquitous and exceedingly luminous natural mineral. Its crystal structure is cubic [31, 39]. Although nominally pure fluorite is colourless, extrinsic impurities within it induce colour [40]. These impurities include Sr and rare earth (RE) elements, which supposedly substitute for Ca [41]. Other studies suggest that the extrinsic impurities exist within CaF_2 as extended defect aggregates rather than as point defects [40, 42]. The RE elements are responsible for the most stand-out luminescence features in photoluminescence, radioluminescence, fluorescence and thermoluminescence of fluorite [39, 40, 42].

The extreme sensitivity of fluorite to stimulation facilitates its use in such areas as dosimetry [43] or for detection of dark matter [44]. The glow curve of fluorite typically has three to six peaks with most of them being of first order kinetics [45–48]. Most measurements on the conventional TL of CaF_2 have been concerned with its kinetics features. However, it is studies based on its TL spectra (e.g [40, 42]) or its PTTL [49–51] that have more effectively aided attempts to understand the physical processes of luminescence in CaF_2.

Studies of PTTL in CaF_2 have been defined by precedence since the earliest ones used UV light to induce PTTL in CaF_2 (e.g. [49, 50]). These are exemplified by Kharita *et al* [52], who studied the effect of annealing on PTTL in natural fluorite using 245 nm illumination; Sono and Mckeever [53], who studied CaF_2:Cu using 307 nm UV exposure; and Sunta [49], who investigated PTTL in green fluorite using a UV source (wavelength unspecified).

The emission of PTTL induced by UV light in CaF_2 is attributed to either hole transfer [49] or electron transfer [41] with defect pairs being involved in both processes. In studies of TL spectra from natural fluorite containing different types of RE impurities, Calderon *et al* [40] concluded that the defects involved are complexes of multiple intrinsic interstitials, vacancies and extrinsic interstitials.

A detailed study in which the PTTL in CaF_2 was induced by 470 nm blue, 525 nm green and 870 nm infrared light and detected between 250 nm and 390 nm has been reported [51]. That work also describes effects caused by illuminating CaF_2 with UV light as well as the influence of illumination temperature on PTTL. The long term behaviour of time-response profiles is analysed using stability theory. The report draws conclusions about deep electron traps as well as a hole trap in CaF_2, their role in PTTL and also looks at the attendant competition effects related to deep electron traps. Mechanisms for the PTTL were discussed. This work forms the basis for most of the presentation that now follows.

5.3.2 Glow curve

Figure 5.18 presents a glow curve of CaF_2 corresponding to 10 Gy beta dose [51]. There are five nominal peaks near 80, 220, 280, 350 and 480 °C denoted I, II, … in that order.

5.3.3 Pulse annealing

Results of the so-called pulse annealing test, which studies the role of electron traps as acceptors or donors in phototransfer, are given in figure 5.19. This example is for PTTL induced by 525 nm illumination for preheating at 10 °C intervals from 50 °C to 600 °C.

The guiding principle in the pulse- or step-annealing procedure is that a glow peak should be stable in intensity unless it is removed in part or fully by preheating or if the preheating depletes the donor for its PTTL. Considering each of the peaks in figure 5.19, we see that peak I first decreases in intensity due to preheating. As the peak reappears under phototransfer, its intensity responds to change at its donors. The latter may be all or any of electron traps of peaks II–V. Although these peaks are stable up to ~300 °C, one of these, namely peak II, weakens in intensity in the same temperature range. Its decrease is coincident with that of peak I. This suggests that the electron trap of peak II is a donor for PTTL at peak I. Peak I then goes on to considerably decrease in intensity between 300 °C and 350 °C, a temperature region in which peaks II and III are removed. This implies that their electron traps are the

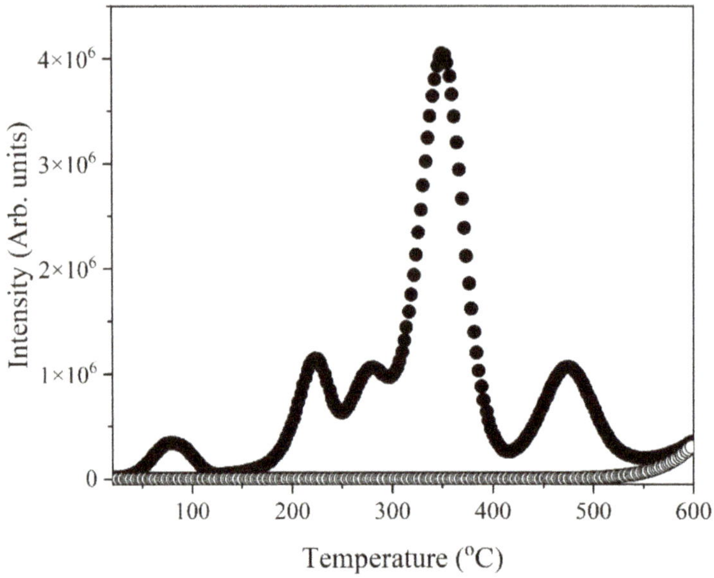

Figure 5.18. A glow curve of CaF_2 measured after beta irradiation to 10 Gy. The background signal is shown for comparison [51].

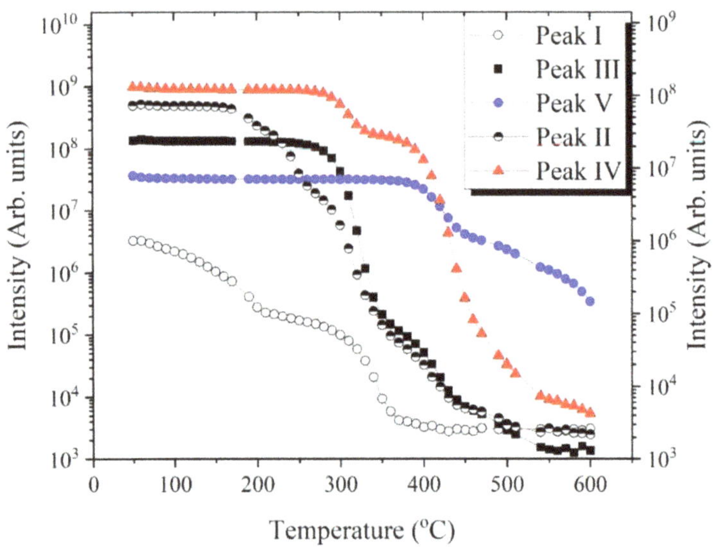

Figure 5.19. The influence of preheating temperature on intensity of peaks I–V. Measurements correspond to illumination by 525 nm light for 100 s each time the irradiated sample is preheated. Reprinted from [51], with the permission of AIP Publishing.

main donors for PTTL at peak I. By similarly matching the change at an acceptor with supposed donors, one concludes that for PTTL at peak II, the donors are the electron traps for peaks III–V, whereas for PTTL recorded at peak III, the source traps correspond to peaks IV and V.

5.3.4 PTTL time-response profiles

The influence of duration of illumination on the PTTL is described in relation to peaks removed by preheating. Doing so offers a means to not only monitor the PTTL but also any intensity change at the supposed donors. The PTTL time-response profiles concerned are shown in figures 5.20(a)–(e).

5.3.4.1 Peak I

Peak I is not regenerated under phototransfer if it is the only one removed by preheating.

5.3.4.2 PTTL after removal of peaks I–II

When peaks I and II are preheated off, they each reappear under phototransfer by either green or blue light. Their putative donors are the electron traps for peaks III–V. The change of intensity of peak I with illumination is shown in figure 5.20(a). Both go through a maximum with illumination. Of note is that when the PTTL is due to illumination by green light, its increase and subsequent decrease is slow. The results for peak II, shown in figure 5.20(b), display similar features [51].

5.3.4.3 PTTL after removal of peaks I–III

Peaks I–III are all regenerated under phototransfer when they are the only ones cleared by preheating. The profile of peak I is shown in figure 5.20(c) and of peak II, in the inset. Although the result of peaks under blue light illumination resembles that in figure 5.20(a) as above, that for peak I under green light again shows a glacial change with illumination.

5.3.4.4 PTTL after removal of peaks I–IV

Figure 5.20(d) displays the time-dependence of the PTTL induced by green light at peaks II and III following removal of peaks I–IV. The plots in the inset display the same behaviour for peaks II, III and IV under blue light illumination. All peaks pass through a maximum in intensity with illumination and are notable for being quite intense.

5.3.4.5 PTTL after removal of peaks I–V

Figure 5.20(e) presents the patterns for peaks IV and V for PTTL induced by blue light. The PTTL is intense and remains so for extended periods of illumination. Interestingly, peak VI, which is not removed by preheating, counterintuitively increases in intensity with illumination (inset). We see this as an instance of competition effects.

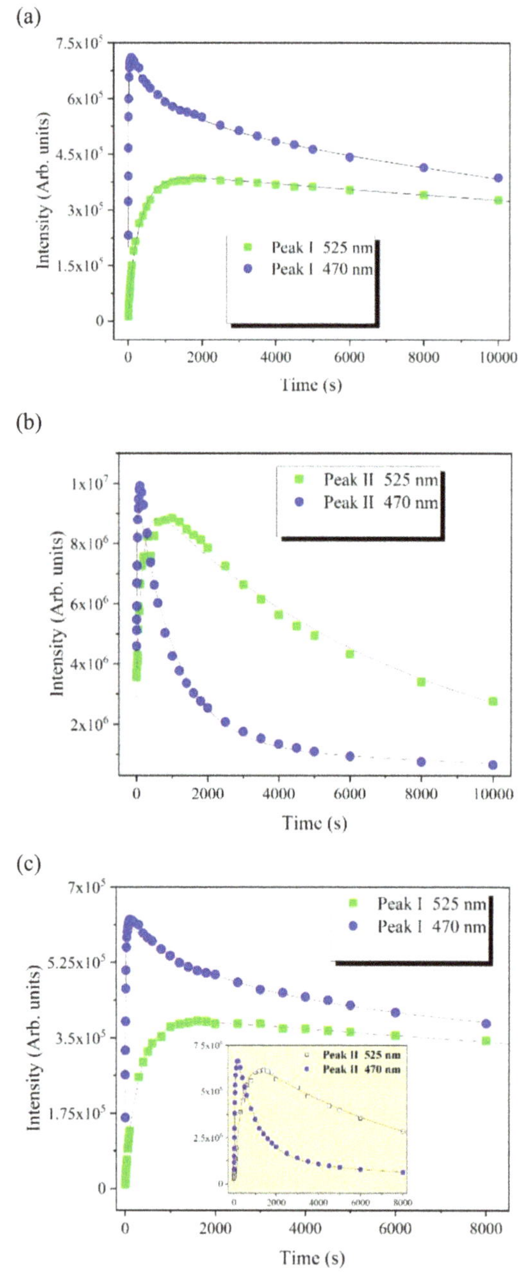

Figure 5.20. The influence of illumination of PTTL intensity for peak I (a) and peak II (b) after preheating to remove only these two peaks. Part (c) displays the results for peaks I and II when preheating removes the first three peaks. The profiles for PTTL after removal of peaks I–IV and I–VI are shown in (d) and (e), respectively. The inset to part (e) is for illumination with infrared light. The lines through the data are mathematical results explained in the text. A plot of ratios of the intensity of the dominant donor to all else in each set corresponding to a given preheating temperature. Reprinted from [51], with the permission of AIP Publishing.

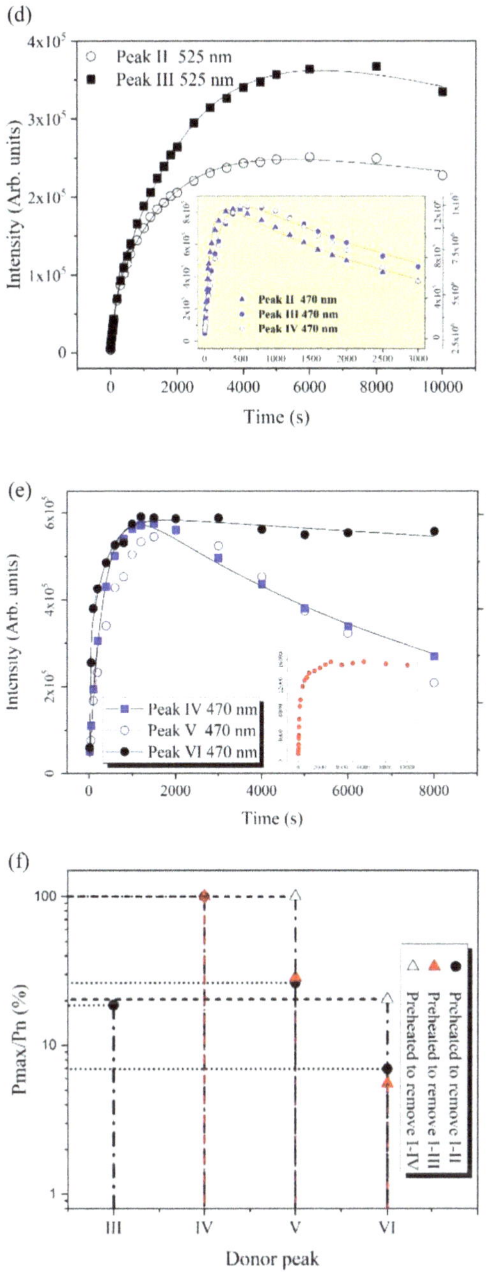

Figure 5.20. (Continued.)

5.3.4.6 Time-response of PTTL induced by infrared light after removal of peaks I–III

PTTL can also be induced by infrared light in CaF_2. Its intensity increases to saturation with illumination [51].

5.3.4.7 Quantifying the role of donor peaks

The contribution of donors corresponding to each preheating can be quantified by monitoring their intensity as a function of the duration of illumination. Figure 5.20(f) is a ratio plot of the most intense donor to all others in a set for PTTL induced by blue light. Such representation aids to empirically decide the proper number of donors for a particular acceptor. This method complements the pulse annealing approach. Here, the PTTL after preheating to remove peaks I–II corresponds to three donors, only two of which are important, hence its system can be regarded as that of one acceptor and two donors. Systems corresponding to preheating that removes up to three peaks consist of one acceptor and two donors, whereas the PTTL measured when four or more peaks are removed have only one donor. In the latter case, the donor is assumed to be a deep electron trap. Similar conclusions apply for PTTL owing to green light [51].

5.3.5 Deep electron traps

5.3.5.1 Phototransfer due to deep electron traps

The possibility of PTTL arising due to phototransfer from deep electron traps is of interest. This is discussed with reference to CaF_2 irradiated not at room temperature but at 500 °C and to a much higher dose of 30 Gy before 470 nm illumination. The dose, which is 10-fold that used for preceding discussions, is intended to fill deep electron traps more effectively in order to facilitate phototransfer from them.

We first look at a one-off measurement on a more massive sample (sample X) than that used thus far. The measurement is intended to highlight features that become prominent when the sample mass is larger. Figure 5.21 (crosses) shows its glow curve following irradiation and illumination. There is a peak near 430 °C. This peak is unlikely to arise due to phototransfer because the duration of illumination used is too short to explain such a large signal. For experimental expediency, follow-up experiments to explore this further were made on the original sample of figure 5.18.

When a glow curve is measured from the latter following irradiation at 500 °C and illumination for longer (figure 5.21), three PTTL peaks appear below 430 °C as indicated. These peaks do not come up if the illumination is omitted (open circles). The peaks must therefore be caused by phototransfer from deep electron traps. Interestingly, peak X still appears in the glow curve obtained without any illumination. This suggests that this and the other three originate from different mechanisms. Chithambo [51] inferred that peak X denotes a hole trap. The reasoning is that peak X appears during measurement of a glow curve due to the recombination of some electrons released at temperatures in the rising edge of peak VI with the trapped holes. The generation of peak X only requires holes to be

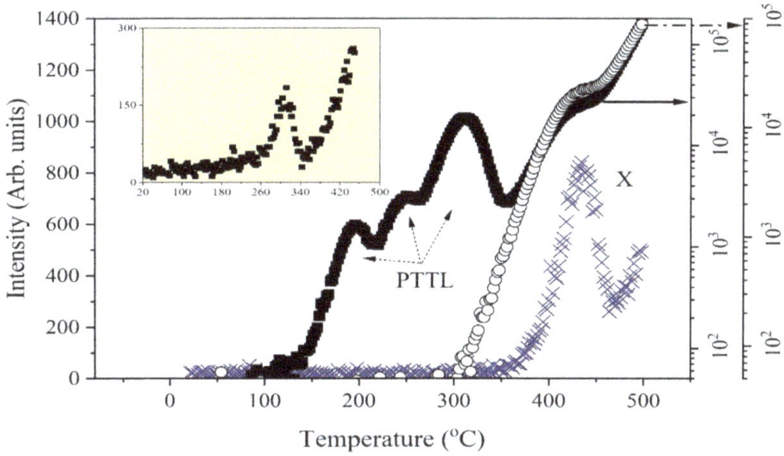

Figure 5.21. Glow curve of sample X after irradiation and illumination for 10 s (crosses). The glow curve of the original sample (of figure 5.19) after irradiation and illumination for 800 s (solid circles) is included together with another obtained after irradiation but without any illumination (open circles). The inset displays a glow curve recorded after irradiation at 500 °C, preheating to 650 °C and illumination by blue light for 1000 s. Reprinted from [51], with the permission of AIP Publishing.

trapped, which explains why illumination is irrelevant for its appearance. The same study [51] shows that CaF_2 also has deep electron traps that activate well beyond 650 °C and influence its PTTL (figure 5.21, inset).

5.3.6 Influence of illumination temperature on PTTL

The influence of illumination temperature on PTTL can be quantified as discussed in section 4.6.8. Since blue or green light is more likely to cause greater loss in luminescence due to stimulation than any change owing to temperature alone, infrared light was used to investigate this aspect in CaF_2 [51]. This approach offers a better means to isolate thermal assistance to optical stimulation in very sensitive materials as is demonstrated elsewhere [54].

The effect of changing the temperature at which the sample is illuminated on the PTTL is discussed for irradiated CaF_2 illuminated at a certain temperature after preheating to remove the first three peaks. This is because a prominent PTTL peak appears only this way following infrared irradiation [51]. The illumination temperature runs from 20 °C to 200 °C and the sample is irradiated anew between measurement to prevent loss of signal due to optical stimulation at each step. A glow curve measured before each irradiation removes the remanent signal. Measurements are repeated with the sequence of illumination temperature reversed from 200 °C to 20 °C. The results of the experiment are displayed in figure 5.22. The profile is independent of whether the measurements are made with the illumination temperature increasing to or decreasing from 200 °C. Thermal assistance during illumination causes the increase in the PTTL intensity. At higher temperatures, thermal

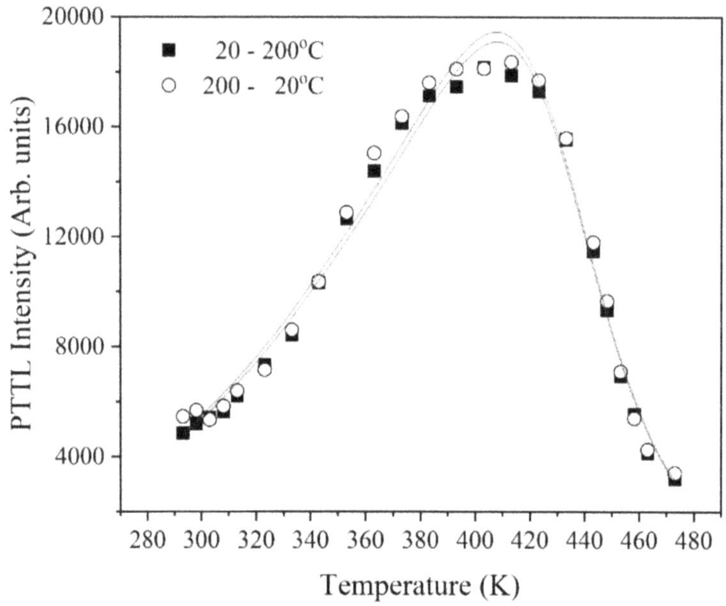

Figure 5.22. The dependence of PTTL intensity on illumination temperature for measurements between 20 °C and 200 °C. Irradiation was done at 500 °C. Reprinted from [51], with the permission of AIP Publishing.

quenching depresses the PTTL. Analysis along the same lines as explained in 4.6.8 leads to $E_a = 0.14 \pm 0.01$ eV, $E_q = 1.1 \pm 0.3$ eV and for the second set (open symbols) $E_a = 0.14 \pm 0.01$ eV, $E_q = 1.1 \pm 0.4$ eV where the activation energies for thermal assistance and thermal quenching are denoted E_a and E_q, respectively.

5.3.7 Role of UV illumination in PTTL from CaF_2

The use of UV sources to induce PTTL from CaF_2 has long been a preferred method (e.g. [49, 50, 52, 53]). This choice, however, raises some questions. While the high power from UV illumination can more effectively probe deep electron traps than blue, green or infrared light can, that advantage is countered by the possibility that UV exposure can also cause ionization. For this reason, interpretation of the resulting signal as PTTL needs re-consideration.

The use of UV light to induce PTTL was examined by illuminating a sample irradiated to 1 Gy [51]. This was done after preheating to remove peak I, I–II, …, I–VI in that order. To restrict the process to phototransfer from very deep electron traps, the sample was preheated to 650 °C. In the resulting glow curves, TL uncharacteristically reappears at all peaks. It is tempting to conclude that unlike blue, green or infrared light, UV illumination causes phototransfer to the electron traps of all five peaks. To isolate the effect of UV exposure only, two glow curves were measured, both without any preheating. One glow curve was recorded after illumination only and the other after beta irradiation only.

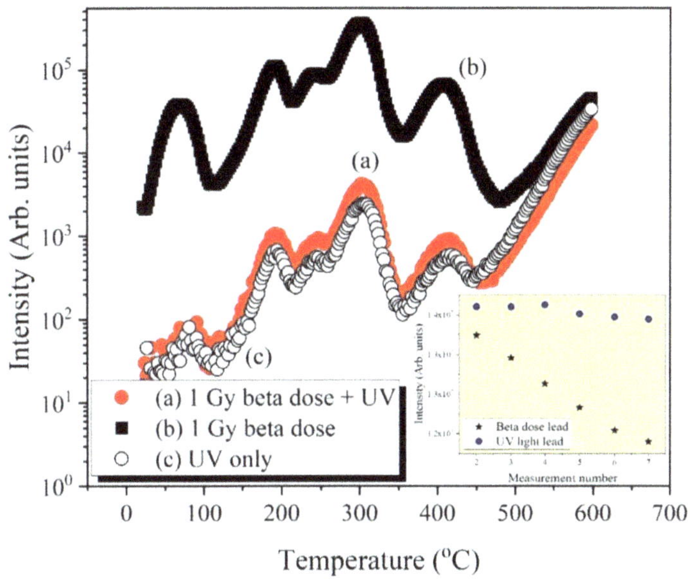

Figure 5.23. Glow curves corresponding to a combination of beta irradiation and UV exposure (a), beta irradiation only (b) and UV illumination only (c). Reprinted from [51], with the permission of AIP Publishing.

These two glow curves are shown in figure 5.23 (and are labelled b and c). Except for intensity, the glow curves show a resemblance. The similarity shows that exposing CaF_2 to UV light produces the same effect as beta irradiation. Therefore, the TL apparent in figure 5.23(c) must mostly be conventional rather than phototransferred TL.

5.3.7.1 Influence of combined beta irradiation and UV exposure on the glow curve

In PTTL protocols, as shown in figure 2.7, irradiation precedes illumination. It is instructive to examine the combined effect of beta irradiation and UV illumination by omitting preheating and sequentially changing over the order of irradiation and illumination. The beta dose is kept constant but the duration of illumination successively increased. The intensities of all peaks are monitored. The inset to figure 5.23 presents the intensity against number of measurement for such an experiment. When the sample is irradiated before illumination, the intensities consistently decrease. This is because UV light reverses the effect of beta irradiation. If, however, illumination precedes irradiation, the succeeding beta irradiation offsets the effect of UV illumination and the change in intensity is minimal. The difference between the two measurements is the signal lost due to illumination. This means that the key effect of UV exposure in this case is to optically remove charge from all peaks. The amount of phototransferred charge is less than that lost due to optical stimulation. This scenario ought to be considered when UV illumination is used to induce phototransfer in CaF_2.

5.3.8 Time-response profiles

PTTL time-response profiles measured from CaF2 are the subject of figure 5.20. These are for peak I (a) and peak II (b) after preheating to remove only this pair. Part (c) shows results for peaks I and II after preheating to remove peaks I–III. The profiles for PTTL obtained after removal of peaks I–IV and I–VI are shown in (d) and (e), respectively. The inset to part (e) is for measurements with infrared light. The lines through the data are results for the same methods leading to equation (3.18). A plot of ratios of the intensity of the dominant donor to all else in each set corresponding to a given preheating temperature is shown in part (f). Such a plot helps one to decide on the proper number of donors. Except for part (d), which corresponds to a single donor, all else can be properly described as a system of one acceptor and two donors. The only exception is peak III, which is affected by competition effects [51].

5.3.9 Stability of donor–acceptor and electron–hole systems

We have thus far modelled systems of an acceptor and donors as autonomous and used their analytical solutions to analyse experimental data as exemplified in figure 5.20. The long term behaviour of the systems under small changes in the initial conditions can be studied by analysing the stability of the solutions. The stability of a simple system of one acceptor and one donor and the application of the theory to study the effect of a hole trap on the concentration of stimulated electrons has previously been discussed [51]. Here we review the latter.

We concluded that UV illumination causes the same effect as beta irradiation. We therefore look at an idealized case where illumination and irradiation are applied simultaneously. The aim of this exercise is to understand the effect of a hole trap on the rate of change of the concentration of stimulated electrons.

We consider a simple system comprising a single hole trap and a single donor electron trap where the instantaneous concentrations of holes and electrons are h and N, respectively. We assume that in the absence of electron traps, the hole concentration increases at a rate $h' = a_2 h$. Similarly, we take it that the concentration of stimulated electrons grows at a rate $n' = a_1 N$ in the absence of a hole trap. When both holes and stimulated electrons are present, their respective numbers will decrease at a rate proportional to the frequency of encounters between them. Therefore, the resulting system can be described as

$$\left.\begin{aligned} N' &= a_1 N - b_1 h N \qquad (a) \\ h' &= a_2 h - b_2 h N \qquad (b) \end{aligned}\right\} \tag{5.14}$$

where a_1, a_2, b_1, $b_2 > 0$ are rate constants. This system has two critical points, namely, (0,0) and $(a_2/b_2, a_1/b_1)$. Since the set of equations in (5.14) are non-linear, we can simplify the procedure by linearizing them in the neighbourhood of a critical point by using the procedure

$$\vec{U'} = \vec{J}\,\vec{U} \tag{5.15}$$

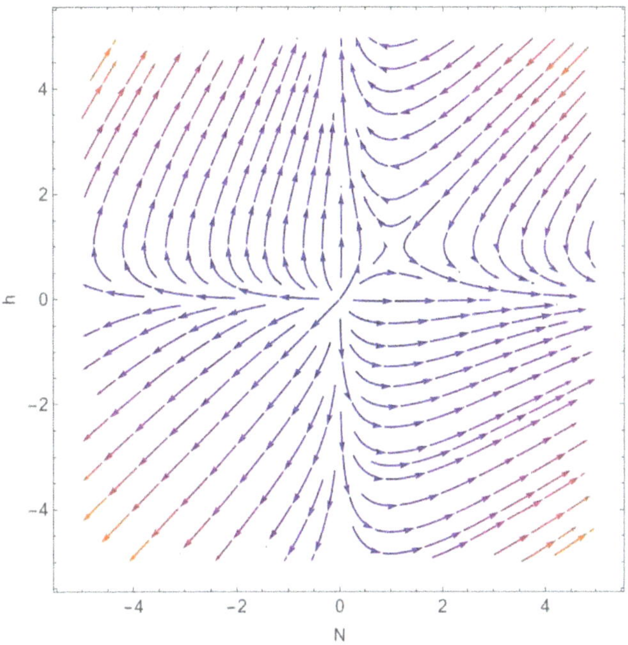

Figure 5.24. The phase portrait for the hole–electron system for which the Jacobian gives two positive eigenvalues. Reprinted from [51], with the permission of AIP Publishing.

where J is the Jacobian and $\vec{U}$ the vector (hN, hN). The system (5.14) will have a solution only if $(\vec{J} - mI)$ is null. The eigenvalues m for $J(0, 0)$ are $m_1, m_2 = a_1, a_2 > 0$. There are two possibilities to consider, namely, either $a_1 = a_2$ or $a_1 > a_2$ (or vice versa). The first possibility predicts an unstable node or a spiral point. The second case refers to an unstable improper node. For the second critical point, $J(a_2/b_2, a_1/b_1)$, $m_1, m_2 = \pm\gamma$, where $\gamma = (a_1a_2)^{1/2}$. Since $a_1, a_2 > 0$ the resulting node should be an unstable saddle point. This predicts that the phase portrait will contain two critical points, one of which is unstable. The resulting phase portraits are shown in figure 5.24. The analysis predicts that the concentrations of holes and electrons cannot mutually annihilate, meaning that there will always be an equilibrium value for both that may be close to but not actually reach zero.

5.3.10 Mechanisms

Although the thermoluminescence of CaF_2 has been extensively studied (e.g. [40–42]), there is no consistent model to explain the phenomena in this material. The spectral features of the TL often reflect characteristic line emissions from RE impurities [40, 42]. It is also well known that CaF_2 acts as a host for many rare earths and their luminescence has been extensively and systematically studied [39]. The common view is that an RE^{3+} cation substitutes divalent Ca in the CaF_2 lattice [40–42, 53]. The charge imbalance is compensated for in various ways including by an O^{2-} at F^- sites [53], or by an F_i^- [41]. These processes respectively produce $(RE^{3+}{-}O^{2-})$ or $(RE^{3+}{-}F_i^-)$

aggregates. Calderon *et al* [40] quoted evidence that the defects cluster into even larger units of several interstitials, vacancies and impurities. Spectral evidence for and implications of defect clustering have been reviewed by Townsend *et al* [55]. The role of other defects such as F and M electron centres and V_k and H hole traps [55] in TL of CaF_2 is not that well understood. Nevertheless, several mechanisms have been proposed to explain PTTL in CaF_2 [51].

It is thought that when heating releases electrons from RE^{2+} ions to produce an excited RE^{3+}, the de-excitation of the latter causes TL. The RE^{2+} ions originate from electron capture at RE^{3+} ions during irradiation. The expectation that TL from this process should reflect emission lines from particular rare earths is realised [40, 42]. In an alternative view, holes become mobile when the temperature is increased. Their recombination with RE^{2+} creates excited trivalent ions by which TL is emitted [41, 53]. The release of holes is assumed to sometimes occur from various modified forms of V_k centres. These two processes are summarised as follows:

$$\left.\begin{array}{ll} RE^{3+} + e^- \rightarrow RE^{2+} \xrightarrow{\text{heat}} RE^{*3+} + h\nu_{\,TL} & (a) \\ RE^{2+} + h^+ \xrightarrow{\text{heat}} RE^{*3+} + h\nu_{\,TL} & (b) \end{array}\right\} \qquad (5.16)$$

To explain PTTL induced by UV illumination in CaF_2, Sunta [49] used the second option on the unit (RE^{3+}–O^{2-}). Here, irradiation causes the transfer of an electron from oxygen to caesium. During UV illumination, a hole is released from the (RE^{3+}–O^{2-}) unit, becomes mobile with temperature, combines with $RE^{2+}$and following this process (process 5.16(b)), the PTTL is emitted. On the other hand, Jain [41] suggested that UV light ionizes photochromic centres releasing an electron leading to the first process summarised in process 5.16(a). Calderon *et al* [40] observes that models based on charge movement through the conduction band or inter-ionic charge transfer are all likely. Proposals for PTTL reported here are therefore particular to the results under discussion.

The glow curve of figure 5.18 is similar to ones where Ce^{3+} is the emission centre (figures 7 and 11 in reference [42]). The PTTL induced by 470, 525 and 870 nm light cannot be due to ionization as is explained for UV light. For this reason, it is assumed [51] that when CaF_2 is irradiated, some resulting free electrons move to RE^{3+} sites converting them to divalent ions. During preheating, electrons are released from some RE^{2+} electron traps and TL is emitted as described earlier. Subsequent illumination then releases electrons, some of which move to available RE^{3+} ions producing RE^{2+}. Since the electron traps do not all have similar coordination, they will activate at different temperatures. However, to account for the glow peak owing to the hole trap (figure 5.21) one can invoke an alternative explanation involving holes. Holes arising from irradiation become mobile. When these recombine at RE^{2+} centres, excited RE^{3+} cations are created. It is their relaxation that produces the relevant peak. In this way, the peak appears whether or not the sample is illuminated or even when the sample is illuminated at temperatures beyond the peak temperature.

5.3.11 Summary

We have described and discussed the PTTL of natural fluorite. In the particular case looked at, the conventional TL glow curve obtained at 1 °C s^{-1} while heating to 600 °C has six peaks, five of which are reproduced under phototransfer. The time-response profiles show various behaviour including the commonly observed increase through a maximum, a drawn-out increase and decrease; or increase towards saturation. The change for the sixth peak, whose intensity counterintuitively increases in intensity with illumination, is suggestive of competition effects. The material has deep electron traps which are also involved in phototransfer. The PTTL intensity scales with illumination temperature due to thermal assistance and is depressed at elevated temperatures owing to thermal quenching. The long term behaviour of time-response profiles as well as the interaction of holes with stimulated electrons have been studied by stability theory. Results show that illuminating CaF_2 with UV light achieves the same effect as irradiating it and the resulting TL is conventional rather than phototransferred. Mechanisms responsible for the PTTL can be discussed in terms of electron transfer from an RE^{2+} centre.

5.4 Calcite

5.4.1 Introduction

Calcite, the most commonly occuring form of natural calcium carbonate ($CaCO_3$), is an intensely thermoluminescent sedimentary mineral with a rhombohedral structure. Calcite is polymorphous to other forms of calcium carbonate such as aragonite [31]. The crystal structure of calcite as rendered in the polyhedral model is shown in figure 5.25. The structure is anisotropic [31] and consists of layers of CO_3 groups

Figure 5.25. A polyhedral model of the crystal structure of calcite. The structure consists of planes of CO_3 groups with interlinkages of Ca atoms between the layers.

with Ca in-between the layers. Each calcium ion is bound to six oxygen atoms. Rare earths and transition metals are some of the major impurities innate to calcite [39]. Of the transition metals, Mn, which substitutes for Ca, is of primary interest for thermoluminescence (TL) of calcite. Indeed, Mn defines the spectral characteristics of some emission bands for TL [56–61], laser-induced time-resolved luminescence spectra [39] or cathodoluminescence [61] of calcite and other carbonates [59, 62]. Townsend *et al* [56] and Calderon *et al* [59] discussed TL spectra measured from natural and synthetic calcite over 200–800 nm with heating from 30 °C to 400 °C. These and other studies (e.g. [63]) show that when the concentration of Mn in calcite is high, emission bands tend to be broad at low temperatures but become narrow with further heating. It is therefore not surprising that narrow bands are characteristic of calcite with only traces of Mn. The broadening of bands is attributed to Mn clusters. The dissociation of these aggregates with temperature transforms the emission bands to narrower line-like structures [56]. Interest in calcite and its various morphologies has been enduring as is exemplified by investigations on its cathodoluminescence under neutron bombardment [64], effect of dose-rate on its luminescence features [65], use of some of its characteristics in computational modelling [66] as well as use of limestone in dating of megalithic limestone buildings to pick but a few examples [67].

Calderon *et al* [58] discussed a model to explain TL in calcite based on measurement of TL spectra and conventional TL on two of its pure habits including Iceland Spar. By distinguishing the spectral emission of Mn from other impurities also intrinsic to calcite, they deduced that the $(CO_3)^{3-}$ anion acts as an electron trap. The same conclusion was reached by de Lima *et al* [68], who also pointed out that $(CO_2)^{-}$ at a different symmetry within calcite may also serve as an electron trap. The question of how electrons transfer from electron traps to recombination centres in calcite has long been somewhat of a puzzle but is in some cases discussed to involve quantum tunnelling (e.g. [4]). Kirsh *et al* [60] invoked tunnelling to account for low temperature glow peaks and suggested that higher temperature peaks arise owing to transport of ionic charge-carriers.

Although the TL of calcite has been extensively studied, reports on its PTTL have been rare and, until recently, consisted only of the report of Lima *et al* [69]. Although of indirect relevance, one can also cite a study on the PTTL of limestone [70]. Lima *et al* [69] investigated PTTL induced by UV illumination in γ-irradiated yellowish natural calcite. Their calcite had three nominal peaks, all of which were regenerated under phototransfer. The dependence of PTTL intensity on duration of illumination extending to 1200 s was suggestive of the long term behaviour of this feature. In particular, the PTTL intensity of the first two peaks seemed to either increase and level off or go through a maximum with illumination. PTTL could also still be observed in cases where all peaks had been preheated off, which implied phototransfer from deep electron traps.

A detailed study on the PTTL of natural calcite induced by 470 nm blue, 525 nm green and 405 nm UV light has been reported by Chithambo [71]. The PTTL is analysed as a system of acceptor and donors with the number of the latter influenced by the preheating temperature. The work examined the effect of illumination temperature on the PTTL,

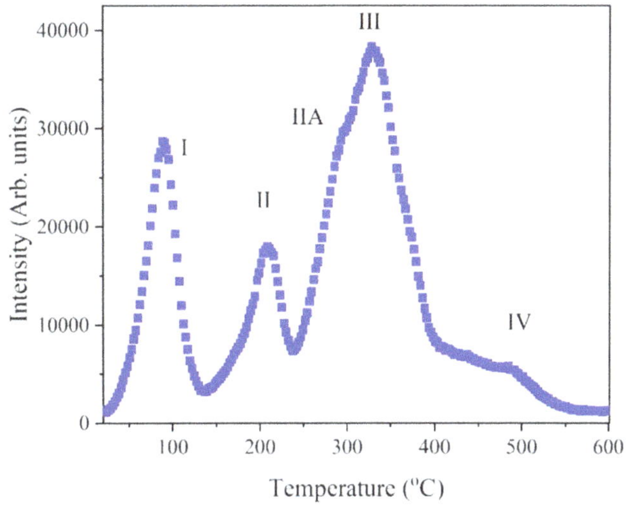

Figure 5.26. Glow curves obtained at 1 °C s^{-1} after irradiation to 119 Gy. There are three nominal peaks near 90 °C, 210 °C and 320 °C (I, II, III) [71]. Reprinted from [71], with the permission of AIP Publishing.

and discussed mechanisms for the PTTL. The PTTL was monitored between 250 nm and 390 nm. We use this study as the presentation that concludes this chapter.

5.4.2 Glow curve

Figure 5.26 is a glow curve corresponding to 119 Gy, the dose used throughout the study. There are three peaks near 90, 210 and 330 °C, denoted I, II and III. The structure of the glow curve shows the possibility of additional peaks. Further examination using 'thermal cleaning' and the T_m–T_{stop} method does show that in addition to peaks I–III, there are two otherwise indistinct peaks near 300 °C and 490 °C [71]. The discussion is concerned with the easily picked out peaks I, II and III.

Some studies (e.g. [72, 73]) are of the view that calcite consists of a continuum of closely spaced peaks. In the context of PTTL, it is impractical to choose preheating temperatures on that basis. Nevertheless, examination using the T_m–T_{stop} method, as shown in figure 5.27, shows that the glow peaks can be treated as discrete. Using the nominal nature of glow peaks, we are concerned with the PTTL of only distinct peaks, that is peaks I, II and III.

5.4.3 Preparatory measurements for PTTL

5.4.3.1 Peaks I–III

Preparatory measurements made after the successive removal of peaks I, II, IIA and III show that PTTL appears following each step. A few relevant but arbitrarily chosen glow curves are shown in figure 5.28. These correspond to preheating to remove the first three peaks and illumination by 470 nm and 525 nm green LEDs for 100 s as well as by a 405 nm laser for 5 s. All peaks in the shaded portion are phototransferred. There is no PTTL when illumination is omitted.

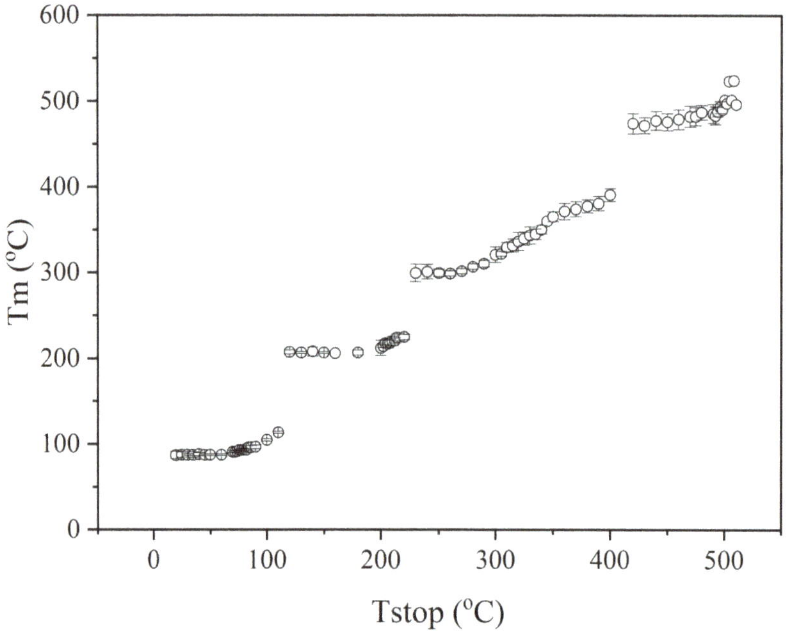

Figure 5.27. The change of peak position Tm with partial heating Tstop. Parts where the peak position is unaffected by the latter each correspond to a distinct peak. Reprinted from [71], with the permission of AIP Publishing.

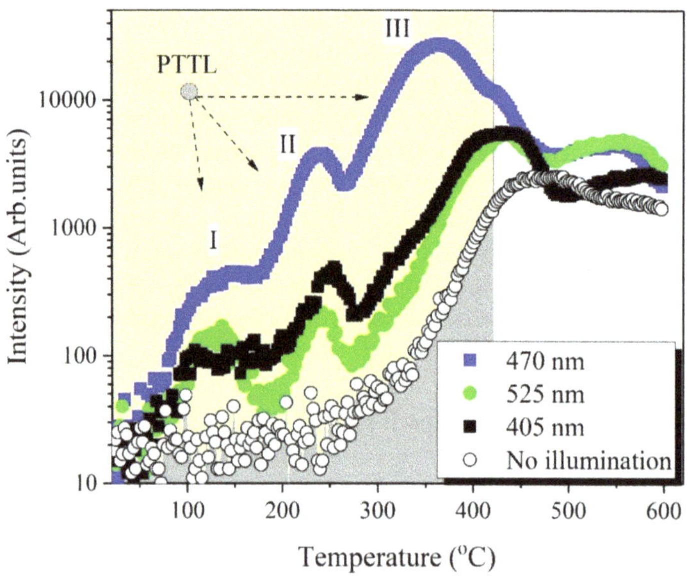

Figure 5.28. Glow curves corresponding to preheating to 430 °C and illumination by 470 nm blue and 525 nm green LEDs as well as by a 405 nm laser [71]. Reprinted from [71], with the permission of AIP Publishing.

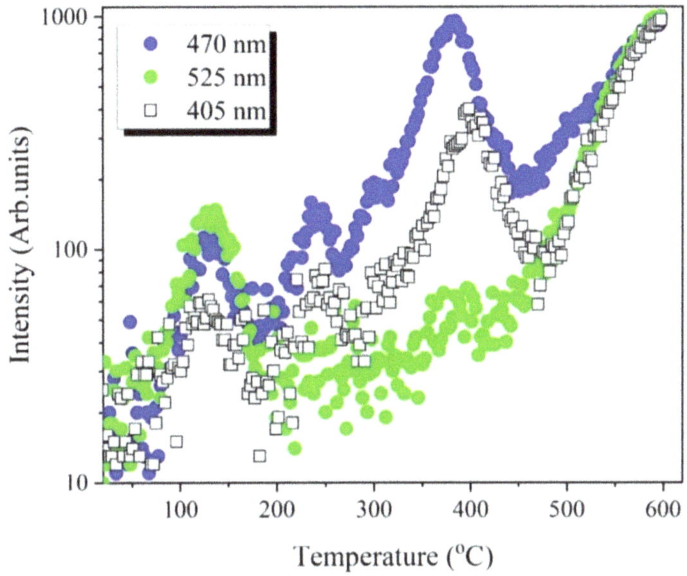

Figure 5.29. PTTL induced from deep electron traps in calcite following preheating to 600 °C. Reprinted from [71], with the permission of AIP Publishing.

5.4.3.2 Phototransfer from deep electron traps

Measurements made following preheating to 600 °C and illumination with 470, 525 and 405 nm light (figure 5.29) show that PTTL can be induced from deep electron traps in calcite.

5.4.3.3 Light-induced fading

Fading of glow peaks due to illumination indicates that such peaks may influence PTTL in some way. In such a study on an irradiated sample of calcite illuminated for various periods up to 30 000 s before measurement of the residual TL each time, all peaks fade with illumination. Thus, the electron traps of all peaks should be involved in the PTTL as acceptors, donors or competitors.

5.4.4 Qualitative association of acceptors and donors

The prediction that electron traps corresponding to all peaks in a glow curve may be involved in phototansfer as acceptors or donors must be experimentally verified. The customary way to do so is to monitor peak intensities following each preheat and illumination in the so-called pulse or step-annealing procedure. Any decrease in intensity of a putative donor should be accompanied by weakening of the PTTL originating from that donor. In that way, one can match an acceptor to a possible donor or donors. We describe measurements where the preheating temperatures were increased from 50 °C to 500 °C at 10 °C intervals and the sample illuminated for 100 s after each preheat using 470, 525 and 405 nm light. The effect of preheating on peak intensity is shown in figure 5.30.

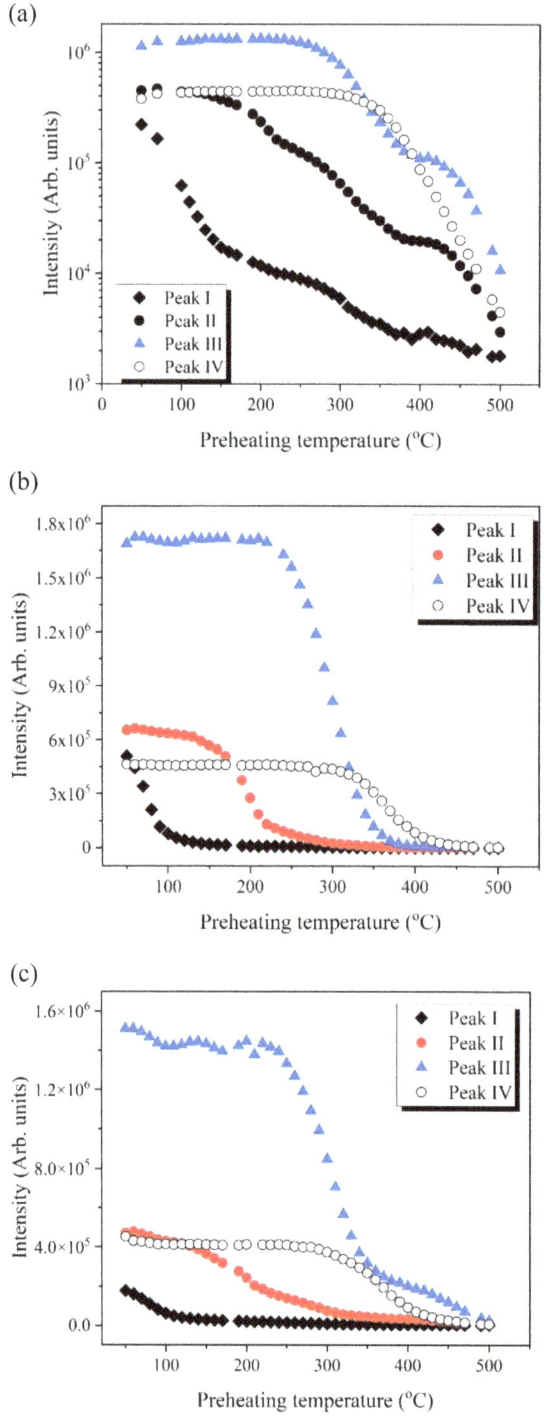

Figure 5.30. Change of intensity with preheating temperature for peaks I and IV for measurements corresponding to 470, 525 and 405 nm in (a), (b) and (c), respectively. Reprinted from [71], with the permission of AIP Publishing.

5.4.4.1 Pulse annealing involving 470 nm illumination

Figure 5.30(a) displays the influence of preheating on the intensity of peaks I–IV. Peak I decreases consistently with preheating temperature. This is caused by the progressive removal of the peak and by depletion of its donor electron traps. Since peaks I and II decrease concurrently, the electron trap of peak II must be the donor for the PTTL at peak I. Using the same reasoning, the PTTL at peak II can be associated with electron traps of peaks III and IV as donors whereas the PTTL at peak III can be partially linked to the electron trap of peak IV as the putative donor.

5.4.4.2 Pulse annealing using 525 nm illumination

The association between a PTTL peak and its supposed donors is more evident in figure 5.30(b) for phototransfer carried out using 525 nm green light. The decrease of peak I matches that of peak II. The electron trap of peak II is therefore likely to be the dominant donor for PTTL observed at peak I. Similarly, the electron traps of peaks III and IV can be linked as donors for PTTL at peak II whereas the electron trap of peak IV can be associated with PTTL at peak III.

5.4.4.3 Pulse annealing using 405 nm illumination

The dependence of peak intensity on preheating temperature for measurements using 405 nm illumination is shown in figure 5.30(c). These plots may be interpreted along the same lines as figures 5.30(a) and (b). However, unlike in measurements corresponding to 525 nm light (figure 5.30(b)) the change for peak III only partially resembles that of peak IV when 405 nm and 470 nm light is used for phototransfer. We deduce that the PTTL at peak III originates from something other than the electron trap of peak IV only. Thus, one can describe the system as consisting of one acceptor and at least two donors with the additional donor being a deep electron trap.

5.4.5 The influence of duration of illumination on PTTL intensity

The influence of duration of illumination is described for PTTL obtained on peaks I–III after preheating to remove peaks I–II; I, II and IIA; I–III and I–IV before illumination. Green light produces PTTL of suitable intensity for study at only some peaks. The aim of measurements after preheating to 600 °C is to monitor PTTL from deep electron traps.

5.4.5.1 PTTL following preheating that removes the first two peaks

When peaks I and II are cleared off first, they are both regenerated under phototransfer by blue or UV light. On the other hand, green light does not induce any PTTL. The dependence of peak intensity on duration of illumination is shown in figure 5.31(a) for peak I and in figure 5.31(b) for peak II. In both cases, the intensity passes through a maximum with illumination. The lines through the data points are mathematical results to be explained later in the text.

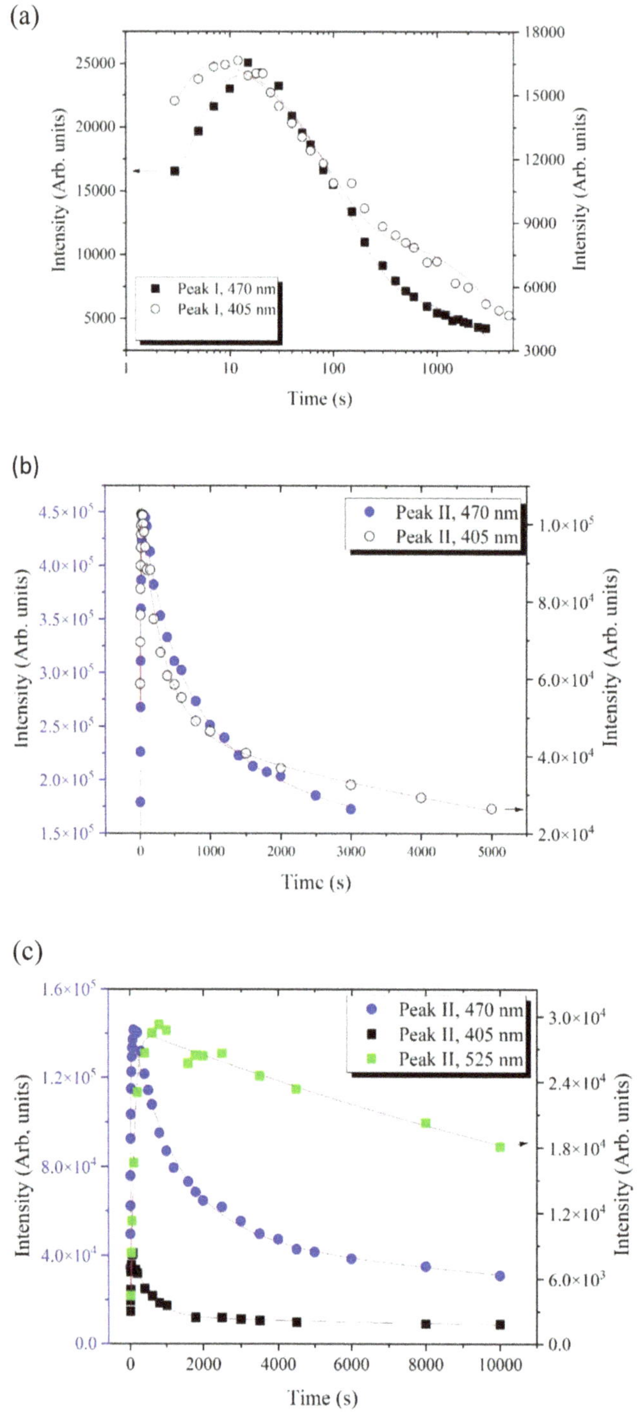

Figure 5.31. Change of intensity with preheating temperature for peaks I and IV for measurements corresponding to 470, 525 and 405 nm in (a) through (f) as explained in the text. Reprinted from [71], with the permission of AIP Publishing.

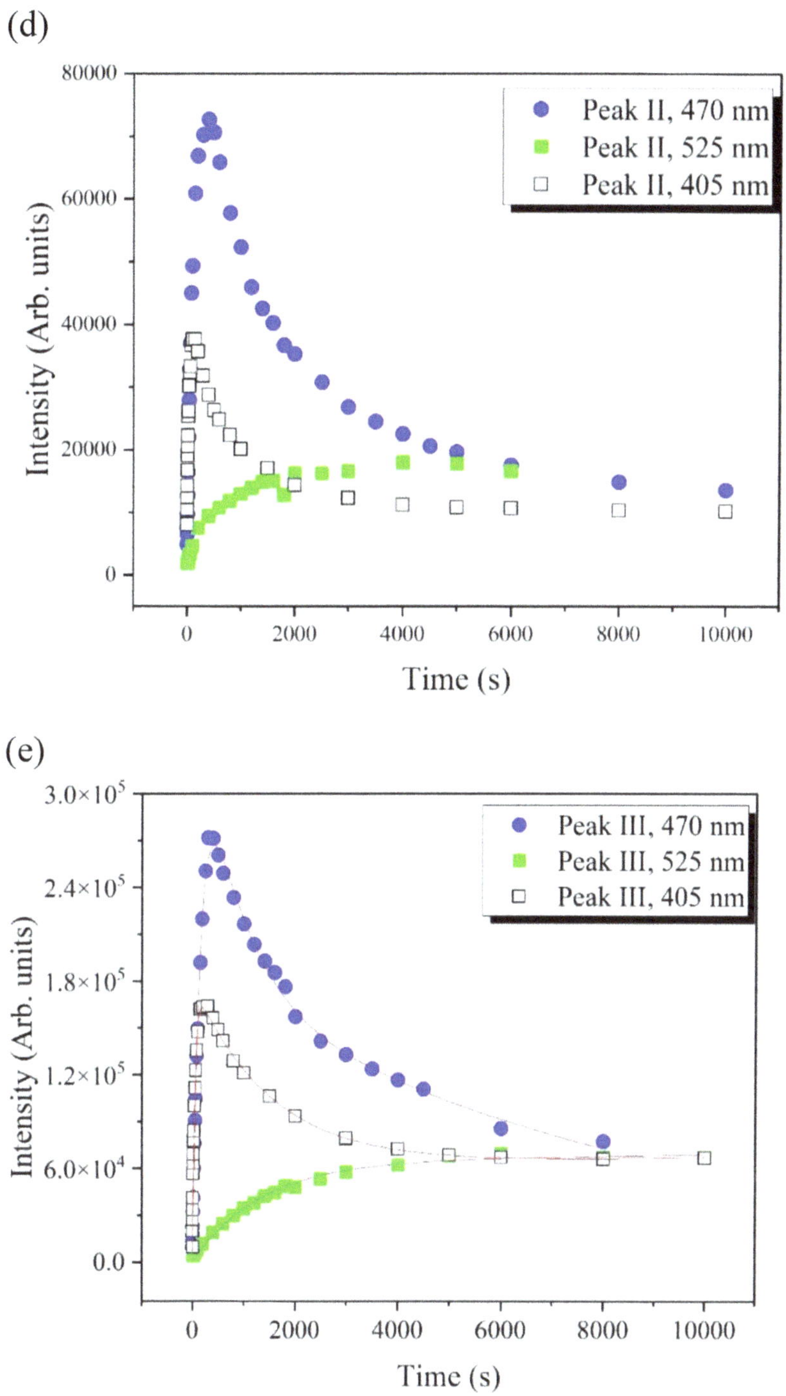

Figure 5.31. (Continued.)

Table 5.1. A summary of systems of acceptor and donor(s) involved in the PTTL discussed. 'DT' denotes the deep electron trap and other symbols are as specified in the text.

Peak label	T_m (°C)	Pulse annealing Donors	Preheating temperature (°C)	Time response Possible donors	Qualitative summary
I	90		140		1A1D
II	210	III, IV	245	III, I	1A2D
IIA	300		320		
III	330	IV	430	IV, DT	1A2D
IV	490		600		

5.4.5.2 PTTL after the removal of peaks I, II and IIA

Following preheating to remove peaks I, II and IIA, only peaks I and II are reproduced under phototransfer. The dependence of PTTL intensity on duration of illumination for peak II corresponding to blue, UV and this time, green light as well, are shown in figure 5.31(c). The intensity following exposure to 405 nm and 470 nm light displays the usual increase through a maximum. In comparison, the PTTL induced by 525 nm stimulation also increases to a maximum but the subsequent decrease is comparatively slow.

5.4.5.3 PTTL after the removal of peaks I–III

When the first three peaks are removed prior to illumination, they are all reproduced under phototransfer by blue, green and UV illumination. Figure 5.31(a) shows the PTTL time-response profile for peak II. The change for peak III is shown in figure 5.31(b). In both results, the PTTL owing to blue and UV illumination increases rapidly to a maximum and decreases thereafter. However, the PTTL stabilises at a level above the initial intensity. This again points to contributions from deeper electron traps. The change of the PTTL due to green light is once again comparatively slow. Table 5.1 summarises systems of acceptor and donor(s) involved in the PTTL as discussed.

5.4.6 Phototransfer from deep electron traps

PTTL measured after preheating to 600 °C offers a means to investigate the contribution of deep electron traps. Figure 5.32 shows selected glow curves measured after various durations of illumination by the more energetic 405 nm light. Each glow curve consists of a single peak near 320 °C, namely, peak III. The peak increases in intensity with illumination. The form of the peak, however, becomes poorly defined with increase in light exposure, particularly at its higher temperature end. The dependence of its PTTL intensity on illumination time is displayed in figure 5.32 and is unusual in that it increases monotonically. By focussing on the change of TL intensity with delay between irradiation and

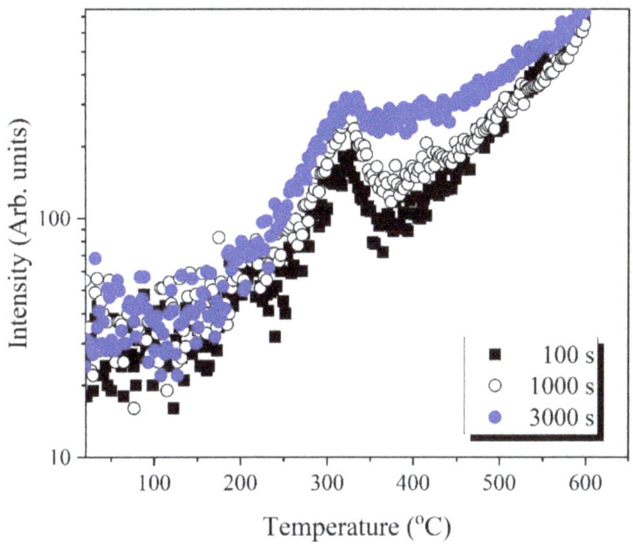

Figure 5.32. PTTL corresponding to preheating to 600 °C and illumination by 405 nm UV light intended to induce transfer from deep electron traps [71]. Reprinted from [71], with the permission of AIP Publishing.

measurement at the higher temperature end of the peak, one notes that although the TL peak fades consistently with delay after irradiation, the region beyond 400 °C does not. The intensity of the ROI between 400 °C and 600 °C first decreases for delays less than 10 000 s or so but increases for longer delays. This is shown in figure 5.32. We conclude that the specious form of figure 5.32 occurs because the increase supersedes any contribution from phototransfer. The cause of the increase is yet unclear but may involve transfer of electrons to concerned electron traps by quantum tunnelling, a well discussed feature for calcite.

5.4.7 Influence of illumination temperature on PTTL

The intensity of optically stimulated luminescence scales with illumination temperature. The intensity usually passes through a maximum with illumination time. The initial increase is ascribed to 'thermal assistance' whereas the subsequent decrease is attributed to frequent incidences of non-radiative transitions or thermal quenching. These two thermodynamically independent thermal effects can be monitored using the methods described in section 3.4.

Figure 5.33 shows the influence of illumination temperature on PTTL intensity. Two peaks, II and III, are regenerated (figure 5.33(a), inset) under phototransfer and their change with stimulation temperature is shown in figure 5.33(a) and figure 5.33(b), respectively. Here, studies were done by illuminating an irradiated sample for 100 s after preheating to remove the first four peaks (peaks I, II, IIA and III) to thereby ensure that phototransfer only ensues from deep electron traps. Phototransfer was induced by 470 nm blue LED and 405 nm UV laser light.

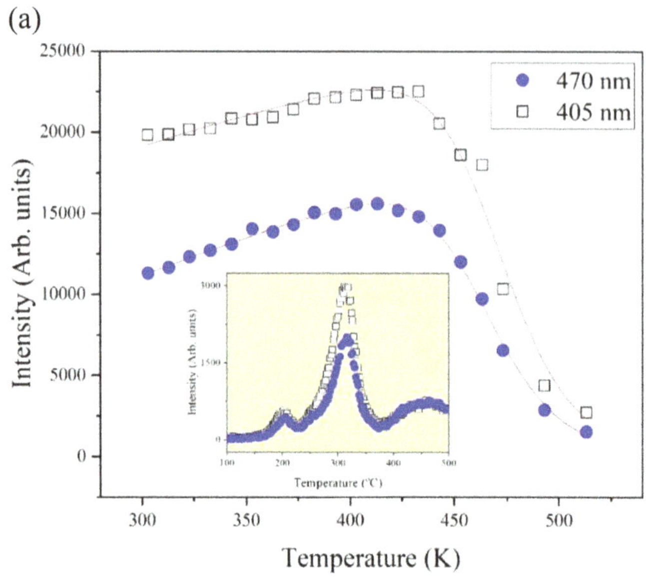

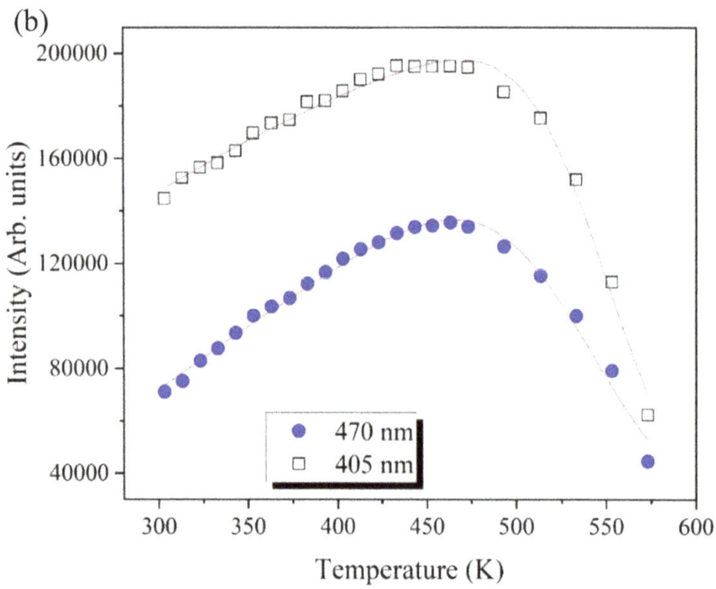

Figure 5.33. The temperature-dependent change in intensity of PTTL at peak II (a) and III (b). The PTTL was induced by 470 nm and 405 nm illumination [71].

The solid lines through the data in figure 5.33 are, in each case, a fit of equation (3.44). The activation energy for thermal assistance corresponding to PTTL at peak II is $E_a = 0.018 \pm 0.004$ eV and 0.035 ± 0.002 eV for measurements with 405 and 470 nm illumination, respectively. The corresponding activation energy for thermal quenching is $E_q = 1.17 \pm 0.63$ eV and 1.11 ± 0.17 eV. In comparison, the values from

Table 5.2. Thermal quenching and thermal assistance activation energy values for calcite.

Peak	Wavelength (nm)	E_a (eV)	E_q (eV)
II	405	0.018 ± 0.004	1.17 ± 0.63
II	470	0.035 ± 0.002	1.11 ± 0.17
III	405	0.023 ± 0.001	1.01 ± 0.12
III	470	0.052 ± 0.002	0.75 ± 0.07

analysis using peak III (figure 5.33(b)) are $E_a = 0.023 \pm 0.001$ eV and 0.052 ±0 .002 eV and $E_q = 1.01 \pm 0.12$ eV and 0.75±0.07 eV for illumination with 405 nm and 470 nm illumination, respectively. The values are summarised in table 5.2.

Thermal quenching in calcite has also been looked at by Kalita and Wary [73] using TL whose glow curve had three peaks. Using each of these, they reported values of the activation energy for thermal quenching as 1.46 ± 0.54, 1.14 ± 0.73 and 1.38 ± 0.40 eV. Kalita and Wary [73] measured their signal using a wide-band filter transmitting from the visible to the infrared. As explained elsewhere [74], values of the activation energy for thermal quenching depend on the emission wavelength, whereas the activation energy for thermal assistance depends on the stimulation wavelength. Values of E_a are consistent with the latter.

5.4.8 Analysis of time-response profiles

The dependence of PTTL intensity on the duration of illumination is analysed using a phenomenological model and by theoretical modelling as explained in chapter 3. The analysis of experimental behaviour based on the phenomenological model is shown as lines through the data in figure 5.30. The PTTL refers to preheating to deplete electron traps of peaks I–II (figure 5.30(a)); I, II and IIA (figure 5.30(b)); and I–III (figure 5.30(c)). In all these examples, the systems can be constrained to comprise one acceptor and two donors, namely, with $j = 2$ in equation (3.18). Although peak I is deduced to have a single dominant donor, its plots (figure 5.30(a)) can in, practice, also only be properly described by a model of two donors.

5.4.8.1 Theoretical modelling

As an illustrative example, we look at the PTTL of peak II induced by 525 nm illumination following preheating that removes the first four peaks (figure 5.30(c)). The charge transport for this system can be described as

$$\frac{dn_{d1}}{dt} = -f_{d1}n_{d1} + n_c(N_{d1} - n_{d1})A_{n1} \tag{5.17}$$

$$\frac{dn_{d2}}{dt} = -f_{d2}n_{d2} + n_c(N_{d2} - n_{d2})A_{n2} \tag{5.18}$$

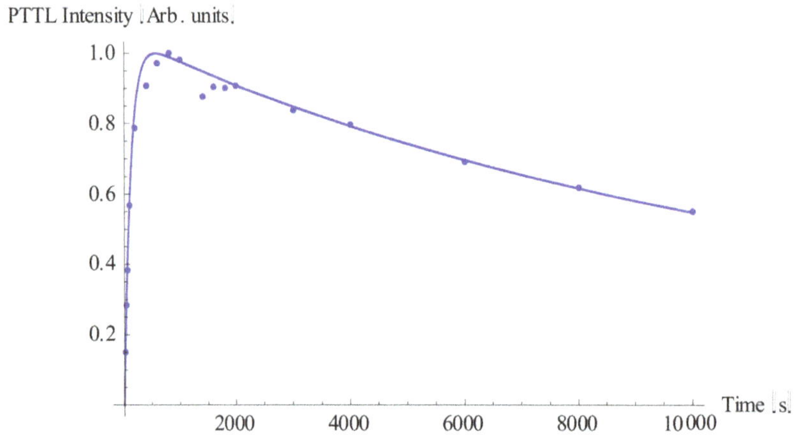

Figure 5.34. Comparison of simulated and experimental results for the time-dependence of PTTL relating to peak II after preheating to 320 °C and illumination by 525 nm green light. The parameters used are $f_{d1} = 8.4 \times 10^{-3}$ s^{-1}, $f_{d2} = 5.5 \times 10^{-6}$ s^{-1}, $N_{d1} = 10^{12}$ cm^{-3}, $N_{d2} = 10^{10}$cm^{-3}, $A_{n1} = A_{n2} = 10^{-19}$ cm^3 s^{-1}, $A_m = 250A_{n1}$ cm^3 s^{-1}. The initial conditions were $N_{d1}(0) = N_{d2}(0) = N_{d3}(0) = 10^{12}$ cm^{-3} ; $n_c = 0$ cm^{-3} [71].

$$\frac{dn_c}{dt} = -A_m h n_c - \frac{dn_{d1}}{dt} - \frac{dn_{d2}}{dt} \tag{5.19}$$

$$\frac{dh}{dt} = \frac{dn_{d1}}{dt} + \frac{dn_{d2}}{dt} + \frac{dn_c}{dt} \tag{5.20}$$

$$I_{\text{PTTL}} = -\frac{dh}{dt} = A_m h n_c \tag{5.21}$$

where N_{di} is the concentration of the ith electron trap, n_{di} the instantaneous electron concentration at the ith electron trap, A_{di} and A_m are the retrapping and recombination probabilities, n_c and h the instantaneous concentration of electrons in the conduction band and holes at the recombination centre and f_{di} the optical stimulation probability. A simulation of the time-response profile is shown in figure 5.34.

5.4.9 Mechanisms

The mechanisms responsible for TL in calcite have long been a matter of research interest (e.g. [3–8]) and have been investigated using a wide range of techniques. Spectral measurements suggest that, apart from intrinsic luminescence, the luminescence centres involve Mn ion impurities [3–8, 29, 30]. TL isometric plots often show a combination of broad and narrow band emissions. The narrow band structures dominate high temperature spectra whereas broad bands are the standout feature at lower temperatures. The broadening of bands is attributed to the clustering of Mn ions whose dissociation with temperature cause line-like structures [3]. The prominent emission near 600 nm is a combination of the transition from the

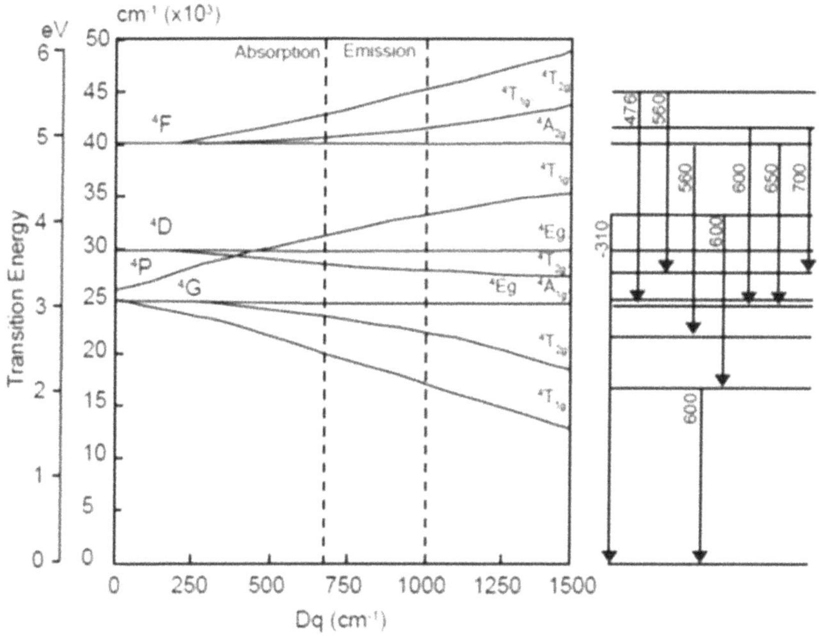

Figure 5.35. Electronic transitions for luminescence in Mn^{2+} in calcite shown alongside the corresponding Orgel diagram as revised by Calderon *et al* [59]. Reprinted from [59], Copyright (1996), with permission from Elsevier.

lowest excited ^4G (T_{1g}) state to the ^6S ground state and other transitions between upper-level excited states as illustrated in figure 5.35 [59].

Chithambo [71] forwarded explanations for the physical processes for the PTTL in calcite on the basis of known processes for TL. Studies using electron spin resonance suggest that the $(CO_3)^{3-}$ anion acts as an electron trap in calcite [5, 15]. The $(CO_3)^{3-}$ centre and Mn^{3+} cations form during irradiation [5]. In the course of heating, some electrons are released from $(CO_3)^{3-}$ centres and move to Mn^{3+} impurities. This results in $(CO_3)^{2-}$ and Mn^{2+} excited states. It is the relaxation of these ions that leads to intrinsic and Mn-impurity related TL emission. We rationalize that once $(CO_3)^{3-}$ electron trapping states form during irradiation, preheating returns some to the $(CO_3)^{2-}$ state. We assume that that these serve as acceptors. When the sample is illuminated after preheating, electrons optically released from other electron traps (donors) convert $(CO_3)^{2-}$ electron traps back to the $(CO_3)^{3-}$ state with some electrons transiting to other Mn impurities. The PTTL monitored thereafter relates to the same process discussed for the TL. These processes are summarised in table 5.3. Roles for other electron traps such as $(CO_2)^{-}$ ions or $(CO_3)^{3-}$ anions are alternative possibilities.

5.4.10 Summary

We have looked at the PTTL induced from natural calcite using 470, 525 and 405 nm illumination. Electronic processes responsible for the PTTL are deduced to

Table 5.3. A summary of processes leading up to emission of PTTL. The PTTL in step 4 may relate to intrinsic or impurity emission.

Procedure	Method TL	PTTL	Result
1. Irradiation	√	√	$(CO_3)^{3-}$ and Mn^{3+} ions form
2. Preheating	N/A	√	$(CO_3)^{3-} + Mn^{3+} \rightarrow (CO_3)^{2-} + Mn^{2+}$
3. Illumination	N/A	√	$(CO_3)^{2-} + Mn^{2+} \rightarrow (CO_3)^{3-} + Mn^{2+} + h\nu_{OSL}$
4. Heating	√	√	$(CO_3)^{2-} + Mn^{2+} + h\nu_{PTTL}$

be related to the relaxation of $(CO_3)^{2-}$ and Mn^{2+} ions. The intensity of the PTTL scales with the measurement temperature.

References

[1] Preusser F, Chithambo M L, Götte T, Martini M, Ramseyer K, Sendezera E J, Susino G J and Wintle A G 2009 Quartz as a natural luminescence dosimeter *Earth Sci. Rev.* **97** 196–226

[2] Bos A J J 2001 High sensitivity thermoluminescence dosimetry *Nucl. Instrum. Meth. B.* **184** 3–28

[3] Schmidt C and Woda C 2019 Quartz thermoluminescence spectra in the high dose range *Phys. and Chem. of Minerals* **46** 861–75

[4] McKeever S W S 1985 *Thermoluminescence of Solids* (Cambridge: Cambridge University Press)

[5] Chen R and McKeever S W S 1997 *Theory of Thermoluminescence and Related Phenomena* (Singapore: World Scientific)

[6] Bailiff I K, Bowman S G E, Mobbs S F and Aitken M J 1977 The phototransfer *technique* and its use in thermoluminescence dating *J. Electron.* **3** 269–80

[7] Schlesinger M 1965 Optical studies of electron and hole trapping levels in quartz *J. Phys. Chem. Solids* **26** 1761–6

[8] Aitken M J 1998 *An Introduction to Optical Dating: The Dating of Quaternary Sediments by the Use of Photon-Stimulated Luminescence* (Oxford: Oxford University Press)

[9] Spooner N A 1994 On the optical dating signal from quartz *Radiat. Meas.* **23** 593–600

[10] Smith B W, Aitken M J, Rhodes E J, Robinson P D and Geldard D M 1986 Optical dating: methodological aspects *Radiat. Prot. Dosim.* **17** 229–33

[11] Wintle A G and Murray A S 1997 The relationship between quartz thermoluminescence, photo-transferred thermoluminescence, and optically stimulated luminescence *Radiat. Meas.* **27** 611–24

[12] Santos A J J, de Lima J F and Valerio M E G 2001 Phototransferred thermoluminescence of quartz *Radiat. Meas.* **33** 427–30

[13] Benny P G and Bhatt B C 2002 High-level gamma dosimetry using phototransferred thermoluminescence in quartz *Appl. Radiat. Isot.* **56** 891–4

[14] Milanovich-Reichhalter I and Vana N 1990 Phototransferred thermoluminescence in quartz *Radiat. Prot. Dosim.* **33** 211–3

[15] Milanovich-Reichhalter I and Vana N 1991 Phototransferred thermoluminescence in quartz annealed at 1000°C *Nucl. Tracks Radiat. Meas.* **18** 67–9

[16] Bertucci M, Veronese I and Cantone M C 2011 Photo-transferred thermoluminescence from deep traps in quartz *Radiat. Meas.* **46** 588–90

[17] Morris M F and McKeever S W S 1994 Optical bleaching studies of quartz *Radiat. Meas.* **23** 323–7

[18] Chithambo M L, Folley D E and Chikwembani S 2019 Phototransferred thermoluminescence from natural quartz annealed at 1000 °C: analysis of time-dependent evolution of intensity and competition effects *J. Lumin.* **216** 116730

[19] Chithambo M L and Dawam R R 2020 Phototransferred thermoluminescence of annealed synthetic quartz: analysis of illumination-time profiles, kinetics and competition effects *Radiat. Meas.* **131** 106236

[20] Chithambo M L, Niyonzima P and Kalita J M 2018 Phototransferred thermoluminescence of synthetic quartz: analysis of illumination-time response curves *J. Lumin.* **198** 146–54

[21] Alexander C S and McKeever S W S 1998 Phototransferred thermoluminescence *Phys. J. Appl. Phys.* **31** 2908–20

[22] Alexander C S, Morris M F and McKeever S W S 1997 The time and wavelength response of phototransferred thermoluminescence in natural and synthetic quartz *Radiat. Meas.* **27** 153–9

[23] Moscovitch M 2011 The principle of phototransferred thermoluminescence *AIP Conf. Proc.* **1345** 323–34

[24] Chithambo M L, Seneza C and Kalita J M 2017 Phototransferred thermoluminescence of α-Al_2O_3:C: Experimental results and empirical models *Radiat. Meas.* **105** 7–16

[25] Folley D E and Chithambo M L 2021 Analysis of thermoluminescence and phosphorescence related to phototransfer in natural quartz *J. Lumin.* **238** 118217

[26] Botter-Jensen L, McKeever S W S and Wintle A G 2003 *Optically Stimulated Luminescence Dosimetry* (Amsterdam: Elsevier)

[27] Galloway R B 2002 Luminescence lifetimes in quartz: dependence on annealing temperature prior to beta irradiation *Radiat. Meas.* **35** 67–77

[28] Chithambo M L and Ogundare F O 2009 Luminescence lifetime components in quartz: influence of irradiation and annealing *Radiat. Meas.* **44** 453–7

[29] Chithambo M L 2018 *An Introduction to Time-Resolved Optically Stimulated Luminescence* (Bristol: Morgan & Claypool Publishers)

[30] Pagonis V, Chithambo M L, Chen R, Chruscinska A, Fasoli M, Li S H, Martini M and Ramseyer K 2014 Thermal dependence of luminescence lifetime and radioluminescence in quartz *Lumin.* J **145** 38–48

[31] Gribble C D 1988 *Rutley's Elements of Mineralogy* (London: Unwin Hyman Ltd)

[32] Harris C, Hlongwane W, Gule N and Scheepers R 2014 Origin of tanzanite and associated gemstone mineralization at Merelani, Tanzania *S. Afr. J. Geol.* **117** 15–30

[33] Olivier B 2006 The geology and petrology of the merelani tanzanite deposit, NE Tanzania *Unpublished PhD Thesis* (University of Stellenbosch, South Africa)

[34] Mills S J, Hatert F, Nickel E H and Ferraris G 2009 The standardisation of mineral group hierarchies: application to recent nomenclature proposals *Eur. J. Mineral.* **21** 1073

[35] Deer W A, Howie R A and Zussman J 1986 *Rock-Forming Minerals—Disilicates and Ring Silicates Vol. IB, The Geological Society* (London: Longman)

[36] Chithambo M L and Folley D E 2020 Dosimetric features, kinetics and mechanisms of thermoluminescence of tanzanite *Physica B: Condens. Matter* **598** 412435

[37] Chithambo M L 2021 Phototransferred thermoluminescence of tanzanite: a matrix-based analysis of time-response profiles and competition effects *J. Lumin.* **234** 117969

[38] Javier Ccallata H, Tomaz Filho L and Watanabe S 2011 Thermoluminescence properties of natural zoisite mineral under γ-irradiations and high temperature annealing *Spectrochim. Acta Part* A **78** 1272–7

[39] Gaft M, Reisfeld R and Panczer G 2005 *Luminescence Spectroscopy of Minerals and Materials* (Berlin: Springer)

[40] Calderon T, Khanlary M-R, Rendell H M and Townsend P D 1992 Luminescence from natural fluorite crystals *Nucl. Tracks Radiat. Meas.* **20** 475

[41] Jain V K 1990 Charge carrier trapping and thermoluminescence in calcium fluoride based phosphors *Radiat. Phys. Chem.* **36** 47

[42] Wang Y, Zhao Y, White D, Finch A A and Townsend P D 2017 Factors controlling the thermoluminescence spectra of rare earth doped calcium fluoride *J. Lumin.* **184** 55

[43] McKeever S W S, Moscovitch M and Townsend P D 1995 *Thermoluminescence Dosimetry Materials* (Nuclear Technology Publishing)

[44] Shimizu Y, Mionwa M, Suganuma W and Inoue Y 2006 Dark matter search experiment with CaF_2(Eu) scintillator at Kamioka Observatory *Phys. Lett.* B **633** 195

[45] Tugay H, Yegingil Z, Dogan T, Nur N and Yazici N 2009 The thermoluminescent properties of natural calcium fluoride for radiation dosimetry *Nucl. Instrum. Meth. Phys. Res.* B **267** 3640

[46] Balogun F A, Ojo J O, Ogundare F O, Fasasi M K and Hussein L A 1999 TL response of a natural fluorite *Radiat. Meas.* **30** 759

[47] Sohrabi M, Abbasisiar F and Jafarizadeh M 1999 Dosimetric characteristics of natural calcium fluoride of Iran *Radiat. Prot. Dosim.* **84** 277

[48] Topaksu M and Yazici A N 2007 The thermoluminescence properties of natural CaF_2 after β-irradiation *Nucl. Instrum. Meth. Phys. Res.* B **264** 293

[49] Sunta C M 1979 Mechanism of phototransfer of thermoluminescence peaks in natural CaF_2 *Phys. Stat. Sol.* **53** 127

[50] Sunta C M 2015 *Unravelling Thermoluminescence* (Berlin: Springer)

[51] Chithambo M L 2022 Phototransferred thermoluminescence of CaF_2: Principles, analytical methods, and mechanisms *J. Appl. Phys.* **132** 055102

[52] Kharita M H, Stokes R and Durrani S A 1995 TL and PTTL in natural fluorite previously irradiated with gamma rays and heavy ions *Radiat. Meas.* **24** 469

[53] Sono D A and McKeever S W S 2002 Phototransferred thermoluminescence for use in UVB dosimetry *Radiat. Prot. Dosim.* **100** 309

[54] Chithambo M L 2021 Thermal assistance in the optically stimulated luminescence of superluminous $Sr_4Al_{14}O_{25}$: Eu^{2+},Dy^{3+} *Physica B: Condens. Matter.* **603** 412722

[55] Townsend P D, Wang Y and McKeever S W S 2021 Spectral evidence for defect clustering: Relevance to radiation dosimetry materials *Radiat. Meas.* **147** 106634

[56] Townsend P D, Luff B J and Wood R A 1994 Mn2+ transitions in the TL emission spectra of calcite *Radiat. Meas.* **23** 433

[57] Khanlary M R and Townsend P D 1991 TL spectra of single crystal and crushed calcite *TL Tracks Radiat Meas.* **18** 29

[58] Calderon T, Aguilar M, Jaque F and Coy-yll R 1984 Thermoluminescence from natural calcites *Physica C: Solid State Phys.* **17** 11

[59] Calderon T, Townsend P D, Beneitez P, Garcia-Guinea J, Millan A, Rendell H M, Tookey A, Urbina M and Wood R A 1996 Crystal field effects on the thermoluminescence of manganese in carbonate lattices *Radiat. Meas.* **26** 719
[60] Kirsh Y, Townsend P D and Shoval S 1987 Local transitions and charge transport in thermoluminescence of calcite *Nucl. Tracks Radiat. Meas.* **13** 115
[61] Diaz M A, Luff B J, Townsend P D and Wirth K R 1991 Temperature dependence of luminescence from zircon, calcite, Iceland spar and apatite *Nucl. Tracks Radiat. Meas.* **18** 45
[62] Chithambo M L, Pagonis V and Ogundare F O 2014 Spectral and kinetic analysis of thermoluminescence from manganiferous carbonatite *J. Lumin.* **145** 180
[63] Down J S, Flower R, Strain J A and Townsend P D 1985 Thermoluminescence emission spectra of calcite and Iceland Spar *Nucl. Tracks* **10** 581
[64] Silletti D K, Brokus S A, Earlywine E B, Borycz J D, Peaslee G F, DeYoung P A, Peters N J, Robertson J D and Buscaglia J B 2012 Radiation-induced cathodoluminescent signatures in calcite *Radiat. Meas.* **47** 3
[65] Ponnusamy V, Ramasamy V, Jose M T and Anandalakshmi K 2012 Effect of annealing on natural calcitic crystals—A thermostimulated luminescence (TSL) study *J. Lumin.* **132** 1063
[66] Chen R, Lawless J L and Pagonis V 2011 A model for explaining the concentration quenching of thermoluminescence *Radiat. Meas.* **46** 1380
[67] Theocaris P S, Liritzis I and Galloway R B 1997 Dating of two Hellenic pyramids by a novel application of thermoluminescence *J. Archaeol. Sci.* **24** 399
[68] de Lima J F, Valerio M E G and Okuno E 2001 Thermally assisted tunneling: An alternative model for the thermoluminescence process in calcite *Phys. Rev.* B **64** 014105
[69] Lima J F, Trzesniak P, Yoshimura E M and Okuno E 1990 *Radiat. Prot. Dosim.* **33** 143
[70] Bruce J, Galloway R B, Harper K and Spink E 1999 Bleaching and phototransfer of thermoluminescence in limestone *Radiat. Meas.* **30** 497
[71] Chithambo M L 2023 Phototransferred thermoluminescence of calcite: Analysis and mechanisms *J. Appl. Phys.* **134** 055109
[72] Guilheiro J M, Tatumi S H, de A, Soares F, Courrol L C, Barbosa R F and Rocca R R 2021 Correlation study between OSL, TL and PL emissions of yellow calcite *J. Lumin.* **233** 117881
[73] Kalita J M, Wary G and Lumin. J 2015 *J.Lumin* **160** 134
[74] Chithambo M L and Kalita J M 2021 *J. Appl. Phys.* **130** 195101

Chapter 6

Other materials of interest

The majority of materials discussed in this chapter share a common property; their glow curves each contain a quasi-continuous distribution of glow peaks. We see three types of such glow curve. The first one consists of collocated peaks; the second, of closely overlapping peaks; and the third kind does not fit neatly into either division. Collocated peaks are those whose components are embedded within one another to such an extent that the product appears single although in reality it is not a *bona fide* single peak. Examples of such behaviour are common in persistent luminescence materials such as $SrAl_2O_4:Eu^{2+}$, Dy^{3+} and have been the subject of a number of studies (e.g. [1, 2]). In the second kind of glow curve, it is possible to just make out the peaks because the rising or falling edges are concealed by adjacent neighbouring peaks. Examples of this are far too numerous to list here but one can single out feldspars [3–7]. Glow curves measured from amorphous materials or disordered solids (e.g. [8]) define the third type because these cannot be described in terms of the same energy band scheme that applies for crystalline materials. These three types of glow curve are of interest because their complexity defies a simple qualitative association of PTTL to well-identified donor peaks. The chapter concludes with a brief look at some selected applications.

6.1 Fluorapatite

Apatite, $Ca_5(PO_4)_3$(F, OH, Cl), is one of the two major varieties of natural phosphates, the other being rock phosphates [9, 10]. The F, OH and Cl-bearing apatites are three types of apatites named fluorapatite, hydroxyapatite and chlorapatite, respectively. The fluorescence of apatite has long been known [11]. Apatite has also drawn interest owing to its luminescence processes, which have been ascribed to quantum tunnelling, anomalous fading and long afterglow [12].

The phototransferred thermoluminescence (PTTL) of natural green fluorapatite was reported by de Farias Soares *et al* [12]. The PTTL was induced by 470 nm light and detected in the UV (250–390 nm) region. The TL glow curve of the apatite

doi:10.1088/978-0-7503-3831-8ch6

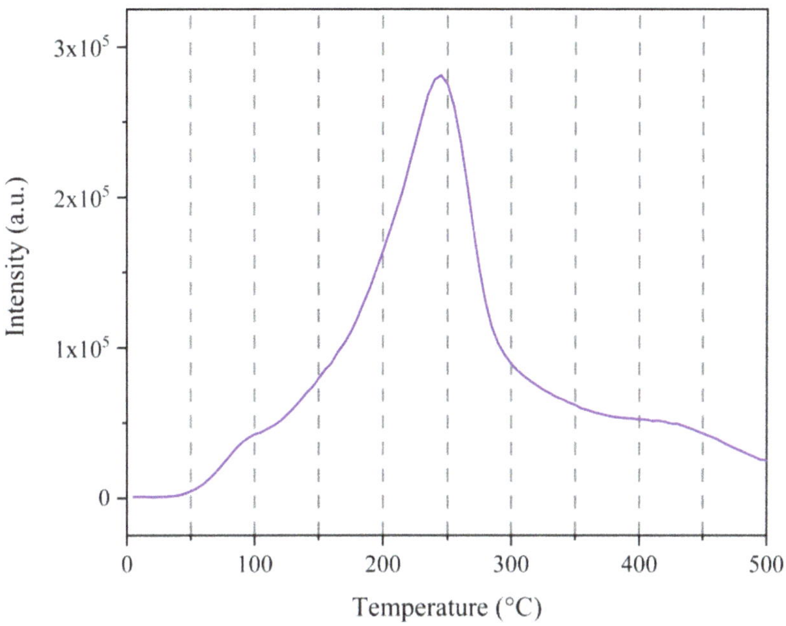

Figure 6.1. A TL glow curve of fluorapatite measured at 5 °C s^{-1} after beta irradiation to 50 Gy. The signal was detected in the UV region (250–390 nm). Reprinted from [12], Copyright (2022), with permission from Elsevier.

studied is shown in figure 6.1. This is of interest because, owing to its form, one cannot readily associate peaks as acceptors and donors.

The glow curve consists of closely overlapping peaks with the prominent one near 250 °C. Using glow curve deconvolution, de Farias Soares *et al* [12] abstracted six component peaks as illustrated in figure 6.2. Of relevance in the PTTL studies are the three last peaks near 250 °C, 350 °C and 450 °C, which we labelled 4, 5 and 6 for reference.

Peak 4 is reproduced under phototransfer at the expense of peaks 5 and 6. This was concluded because a decrease in intensity of the region of peaks 5 and 6 with preheating caused the PTTL to weaken. The same PTTL peak was also observed following preheating to 500 °C and illumination at 170 °C (and other temperatures between 50 °C and 250 °C). The PTTL was deduced to originate from deep electron traps as is customary with such a result. The appearance of PTTL after illumination at high temperatures is thought to exemplify the role of thermal assistance in phototransfer (figure 6.3).

6.2 Obsidian

Obsidian is formed of cooled lava. The extruded lava can reasonably be expected to be an admixture of various molten minerals. The rapid cooling inhibits crystal formation which means that, with its disordered form, obsidian is glass. Obsidian has been used as a TL dosimeter [13, 14], and its spectral characteristics used as a means to re-examine

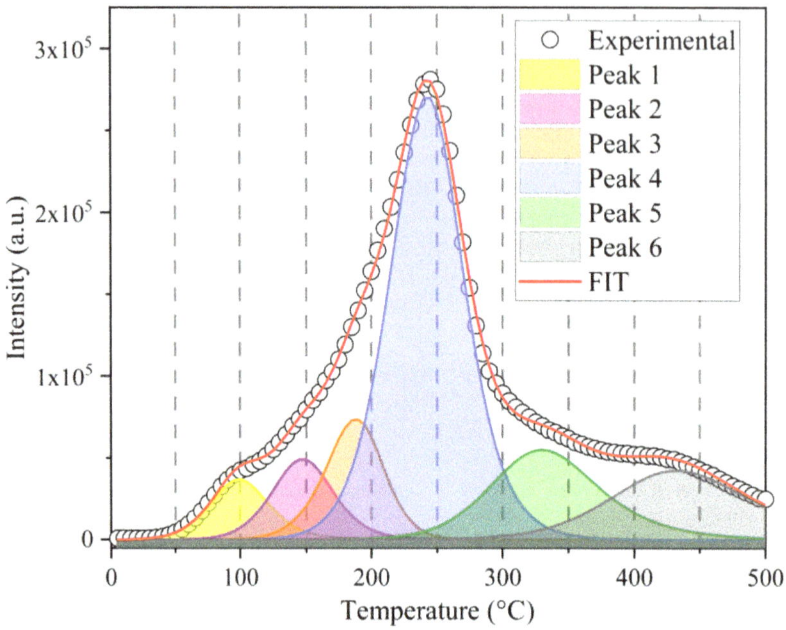

Figure 6.2. The glow curve of fluorapatite corresponding to 50 Gy irradiation overlain with its components as obtained using deconvolution [12]. Reprinted from [12], Copyright (2022), with permission from Elsevier.

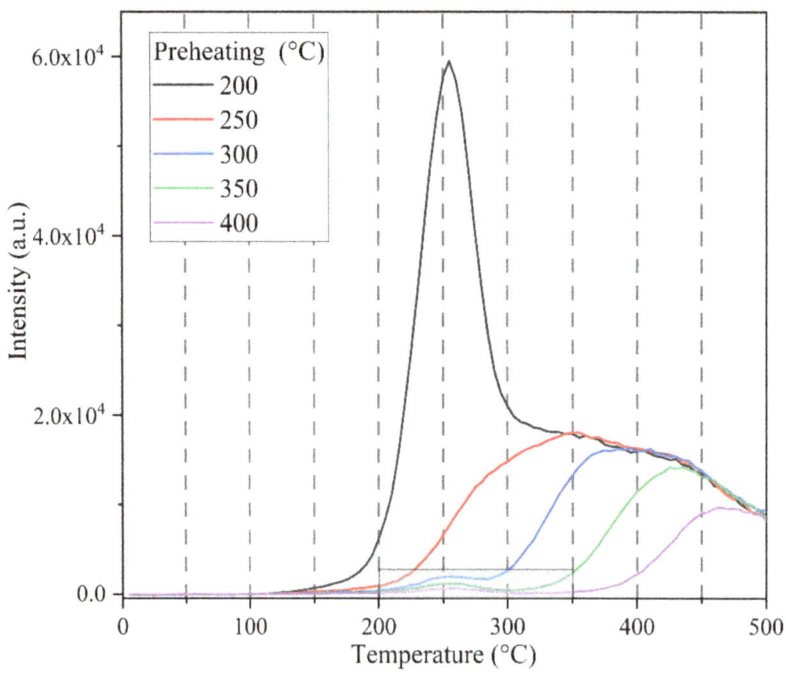

Figure 6.3. The peak near 250 °C reappears under phototransfer and its intensity is related to various preheating temperatures. Reprinted from [12], Copyright (2022), with permission from Elsevier.

its use for cross-calibration of laboratory radiation sources [14]. Studies of PTTL of obsidian are rare and the report of Alzahrani *et al* [15] is relevant. The sample they studied was obsidian SiO_2 (SiO_2 75%; MgO 20%; other material 5%). The PTTL was induced from γ-irradiated obsidian using a 254 nm UV source. A PTTL peak observed at 75 °C after preheating to 500 °C was attributed to deep electron traps.

6.3 $CaSO_4$: Mg

$CaSO_4$ is a dosimetry material with a sensitivity that well exceeds that of LiF-TLD 100, the industry standard [16]. The TL properties of $CaSO_4$ and several of its doped variants have been extensively studied [16] and its appeal persists (e.g. [17]). The TL glow curve tends to be complex consisting of closely overlapping peaks. The PTTL of $CaSO_4$:Mg induced by 470 nm blue light was studied by Guckan *et al* [18]. The emission was detected in the UV region on beta irradiated samples. Their glow curve is shown in figure 6.4(left). Approximate positions as identified by Guckan *et al* [18] are shown. As a means to associate acceptor peaks to donor ones, the glow curve was resolved into seven components (peaks I–VII) as is shown in figure 6.4(right).

Preliminary measurements in which the sample was preheated in turn to remove each of peaks I through VII, showed that PTTL could only ensue when the first four peaks had been removed. An example of the PTTL peak and a series of related glow curves is shown in figure 6.5. When the sample was preheated to 280 °C (to remove peaks I–IV) prior to illumination at various temperatures between room temperature and 220 °C, a PTTL peak was observed near 180 °C. This behaviour resembles that observed in synthetic quartz [19] where PTTL only appears after a series of peaks are first cleared. The study also showed that PTTL intensity in $CaSO_4$: Mg scales with illumination temperature (their figure 6.4). The dose response of the PTTL turned out to be linear for PTTL measured at room temperature or at 100 °C. The result was adjudged as pointing to the possibility of the PTTL being used for dose reassessment.

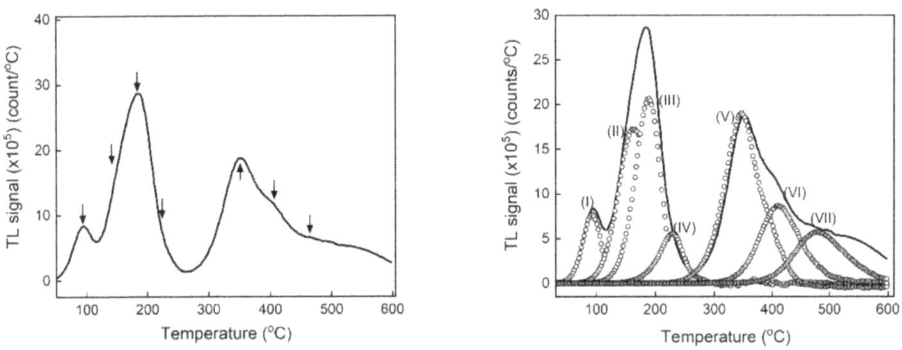

Figure 6.4. A glow curve of $CaSO_4$:Mg measured at 5 °C s^{-1} after beta irradiation to 100 Gy. The approximate position of peaks identified are shown (left). Result of thermal cleaning showing up to seven component peaks at 90, 160, 190, 230, 350, 410 and 480 °C (right). Reprinted from [18], Copyright (2019), with permission from Elsevier.

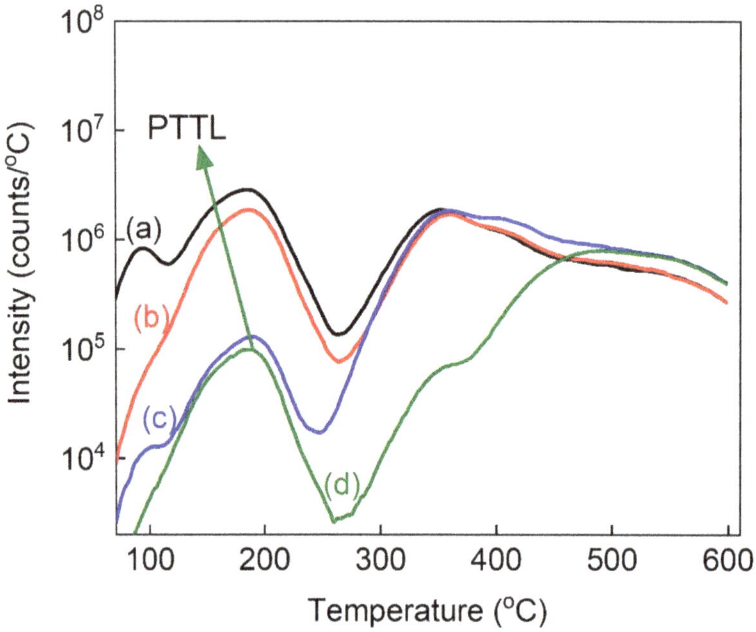

Figure 6.5. A comparison of glow curves measured without any preheating (a) following preheating to intended to remove peak I, and illumination (b) after preheating to 220 °C and illumination where evidence of phototransfer is apparent (c). When the heating is extended to 265 °C and the sample illuminated, a clear PTTL peak emerges as labelled (d). Reprinted from [18], Copyright (2019), with permission from Elsevier.

6.4 KCl

Alkali halides, for example, as variants of LiF have long been used for dosimetry because of their high sensitivity to thermal stimulation [16, 22]. Indeed, the first study on PTTL conducted by Stoddard [20] was on NaCl. Here we single out a study on features of PTTL in $KCl:Eu^{2+}$ as reported by Pedro-Montero *et al* [22]. The glow curve of their x-ray irradiated sample measured at 5 °C s^{-1} showed three peaks at 97, 117 and 197 °C (say peaks 1, 2 and 3). The study monitored PTTL induced by light of wavelengths in the range 450–700 nm. For a particular example where 560 nm was used, the PTTL at 97 °C and 117 °C was attributed to the electron trap of the peak at 197 °C. Pedro-Montero *et al* [22] reasoned that the glow peak at 197 °C is associated with emission at an F centre whereas those at 117 and 197 °C correspond to F_Z centres (ordinarily a halogen ion vacancy with a neighbouring divalent alkali-metal impurity).

6.5 Microcline

6.5.1 Introduction

Microcline ($KAlSi_3O_8$) is a common potassium rich feldspar. It forms in various feldspar-rich rocks including granite or granodiorite and is also found in metamorphic rocks such as gneisses and schists [24, 25]. TL glow curves of microcline are complex and show multiple overlapping peaks due to the presence of a continuum distribution of electron traps [3–5, 26–29]. The activation energies of the electron traps span the range between 0.40 eV and 1.80 eV as the following several examples show. Strickertsson [25] reported values between 0.76 eV and 1.80 eV; Visocekas *et al* [26] gave values between 0.40 eV and 1.4 eV; and Chruścińska [27] had 1.00–1.70 eV. Several other studies (e.g. Pagonis *et al* [28] or Polymeris *et al* [4, 29]) show an electron trap distribution consistent with these results. Kalita and Chithambo [3] reported a continuum distribution of electron traps in microcline with values of activation energy lying between 0.71 eV and 1.15 eV.

Optically stimulated luminescence (OSL) and infrared light stimulated luminescence (IRSL) of various feldspars have likewise been studied and exploited for dosimetric and dating applications. Sfampa *et al* [5] studied ten K-feldspars belonging to three different species of feldspar consisting of sanidine, orthoclase and microcline. They described their experimental results by analytical expressions based on a tunnelling recombination mechanism outlined elsewhere [6] and found a correlation between the TL, OSL and IRSL properties of various feldspars.

The notion of electron transport between point defects in materials through quantum tunnelling was initially discussed by Tachiya and Mozumder [30, 31]. Later work (e.g. [27, 32]) discussed the evidence of quantum tunnelling in feldspar. In a relevant later study on the effect of optical bleaching by blue and infrared light on its TL glow curves, Kalita and Chithambo [33] noticed that the decay curves obtained under 470 nm blue light and 870 nm IR light stimulation differ. Whereas the decay times of blue light stimulated luminescence are affected by dose, those corresponding to infrared light are not. The emission due to blue light is linked to transitions from localized energy levels to the conduction band whereas the IRSL emission is concluded to reflect transitions between localized levels only [33].

The TL, OSL and IRSL of microcline have been studied extensively and their features are well documented [34]. On the other hand, there are comparatively fewer studies on its PTTL. In this regard, Robertson *et al* [35] looked at the effect of bleaching of TL in feldspars by UV (322 nm and 370 nm) illumination as well as by 500 nm green light. Their feldspars showed evidence of phototransfer but the degree of charge transfer was found to depend on the wavelength of light used.

Kalita and Chithambo [36] studied the PTTL of microcline induced by 470 nm blue and 870 nm infrared light. The PTTL was detected in the 250–390 nm (FWHM) UV region and induced using 470 nm blue and 870 nm infrared light. Retaining labels in the original study (i.e. [36]), the PTTL obtained after illumination by blue light will be labelled B-PTTL and that after illumination by infrared light will be referred to as IR-PTTL.

6.5.2 Characteristics of TL and PTTL glow curves

Figure 6.6(a) shows a conventional TL glow curve measured at 1 °C s^{-1} after irradiation to 40 Gy. There are no obvious peaks in the glow curve. However, one can pick out five maxima apparent at 90, 123, 166, 298 and 391 °C, labelled P1–P5. These glow peaks are

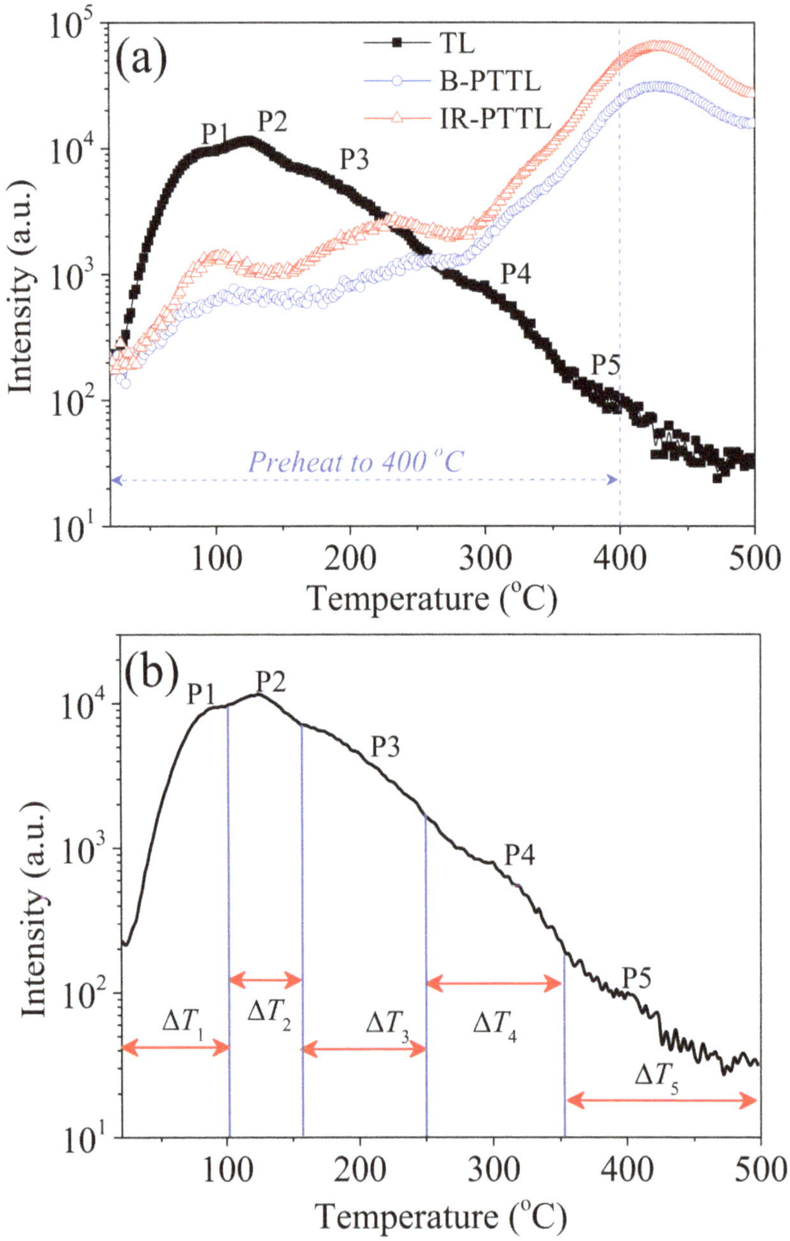

Figure 6.6. A TL glow curve of microcline recorded at 1 °C s^{-1} following irradiation compared with ones obtained after illumination by blue light (B-PTTL) and infrared light (IR-PTTL). All glow curves were measured after irradiation to 40 Gy following preheating to 400 °C and illumination for 100 s (a). Segments of a conventional TL glow curve where ΔT_1, ΔT_2, ΔT_3, ΔT_4 and ΔT_5 are temperature ranges which subsume peaks P1, P2, P3, P4 and P5, respectively. Reprinted from [36], Copyright (2022), with permission from Elsevier.

components of several closely overlapping components [36]. With such a glow curve, it is impractical to select preheating temperature on the need to remove specific peaks. Thus, as a preparatory test, the sample was irradiated to 40 Gy and preheated to an arbitrarily chosen temperature of 400 °C. The sample was then illuminated by either blue or infrared light and heated thereafter to monitor PTTL. The resulting glow curves are shown in figure 6.6(a) where PTTL does ensue following blue or infrared light illumination. However, the PTTL peaks are as ill-defined as they are in the corresponding conventional TL glow curves. Two other notable results are that the portion beyond 400 °C increases in intensity over that of the glow curve when the sample is illuminated by either light source and that the intensity of the IR-PTTL peaks exceeds that of the B-PTTL peaks.

If a glow curve shows discrete peaks, the characteristics of individual PTTL glow peaks reproduced after illumination of a preheated sample can be more easily explored. The microcline in question here has, however, a continuous distribution of electron traps with activation energy between 0.71 eV and 1.15 eV [36]. As such, the peak maxima in figure 6.6(a) may each reflect a combination of several components. Since there are no clear peaks to work with, the PTTL is examined in terms of regions of interest as illustrated in figure 6.6(b). These cover the regions as illustrated and so chosen to include peaks P1, P2, P3, P4 and P5. In this way, the donor and acceptor traps involved in the phototransfer could be studied.

6.5.3 Step-annealing

Figure 6.7 shows the results of pulse annealing measurements intended to distinguish donors and acceptors involved in the phototransfer process. The measurements were made between 60 °C and 500 °C at 10 °C intervals following illumination by blue or infrared light for 100 s each time. The intensity was monitored as the area under each region of interest. Figure 6.7(a) shows examples of glow curves obtained after preheating to specific temperatures and illumination by blue light. Figure 6.7(b) shows similar results for illumination by infrared light. The preheating temperature affects the intensity of the PTTL peaks. The change of the intensity of each peak (segment) against preheating temperature for illumination by blue light is shown in figure 6.7(c) and for illumination by infrared light in figure 6.7(d).

Some useful conclusions can be drawn by matching the changes of PTTL peaks with those of supposed donors. Following the interpretation of pulse annealing results as has been done several times in the preceding chapters, we conclude that the electron trap of peak P2 serves as the main donor for PTTL peak P1. In the same way, the features of figure 6.7 suggest that the electron traps of peaks P3 and P4 act as the main donors for PTTL peak 2 whereas P5 acts as a weak donor for the same PTTL. Similarly, the donor peaks of P3 are P4 and P5 with P4 as the main donor; and the donor peak of P4 is P5 [36]. For this microcline, preheating to 500 °C or beyond does not produce any PTTL.

6.5.4 PTTL time-response profiles

The dependence of PTTL intensity on duration of blue light illumination is shown in figures 6.8(a)–(d) for peaks P1, P2, P3 and P4. The profiles are archetypal and all go

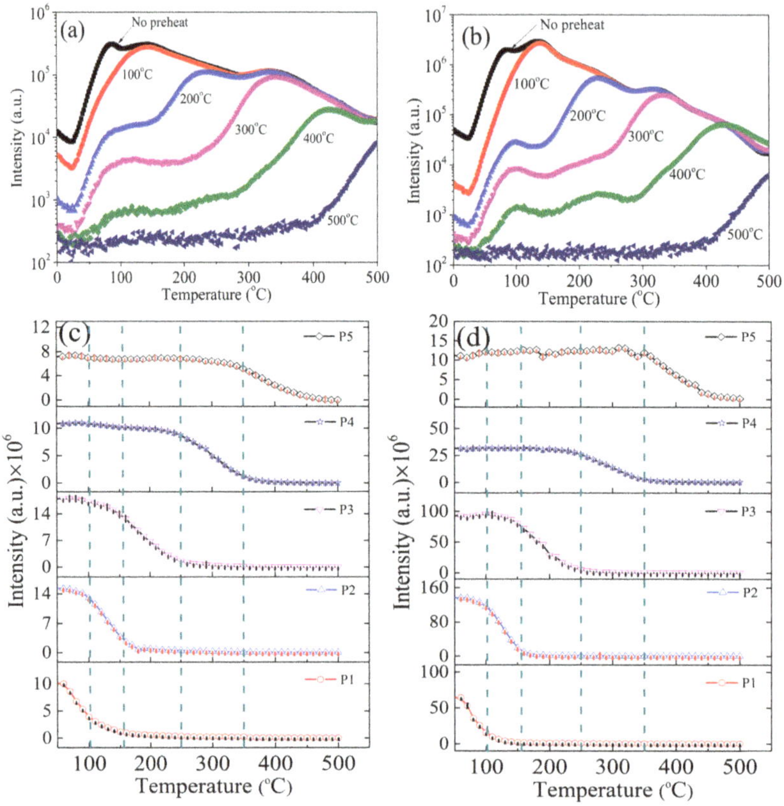

Figure 6.7. A selection of glow curves of microcline measured after preheating to specific temperatures and illumination for 100 s by (a) blue light and (b) infrared light. The change of intensity of peaks P1–P5 against preheating temperature for measurements corresponding to (c) blue light and (d) infrared light. The dotted lines delineate regions of interest used for peaks P1, P2, P3, P4 and P5. All measurements correspond to 40 Gy. Reprinted from [36], Copyright (2022), with permission from Elsevier.

through a maximum with illumination time. The intensity of the supposed donor peak P5 in (e) decreases with time as might be expected. Similar results appear for illumination with infrared light [36].

Previous studies (e.g. [5]) show that the glow peaks in microcline are composite. The nature of the TL and PTTL presented in figure 6.6(a) are suggestive of similar behaviour. In this way, the nature of the PTTL of microcline is similar to that of tanzanite [37] in that it can be analysed in terms of a system of one acceptor and an indefinite number of donors. For any given acceptor, we model this as a system of one acceptor, labelled k, and n donors. The concentration of electrons in the electron trap corresponding to the acceptor is labelled N_k and those at the donor electron traps N_j where $j = k;1\ldots, n$ is an integer. We assume that during illumination, electrons are stimulated from a donor electron trap at a rate $f = \Phi\sigma$ where Φ is the incident photon flux and σ the photoionization cross-section. If the number of electrons stimulated from the ith donor is $f_i N_i$ then the portion captured at an acceptor is $\alpha_i f_i N_i$ where α_i is a proportionality constant.

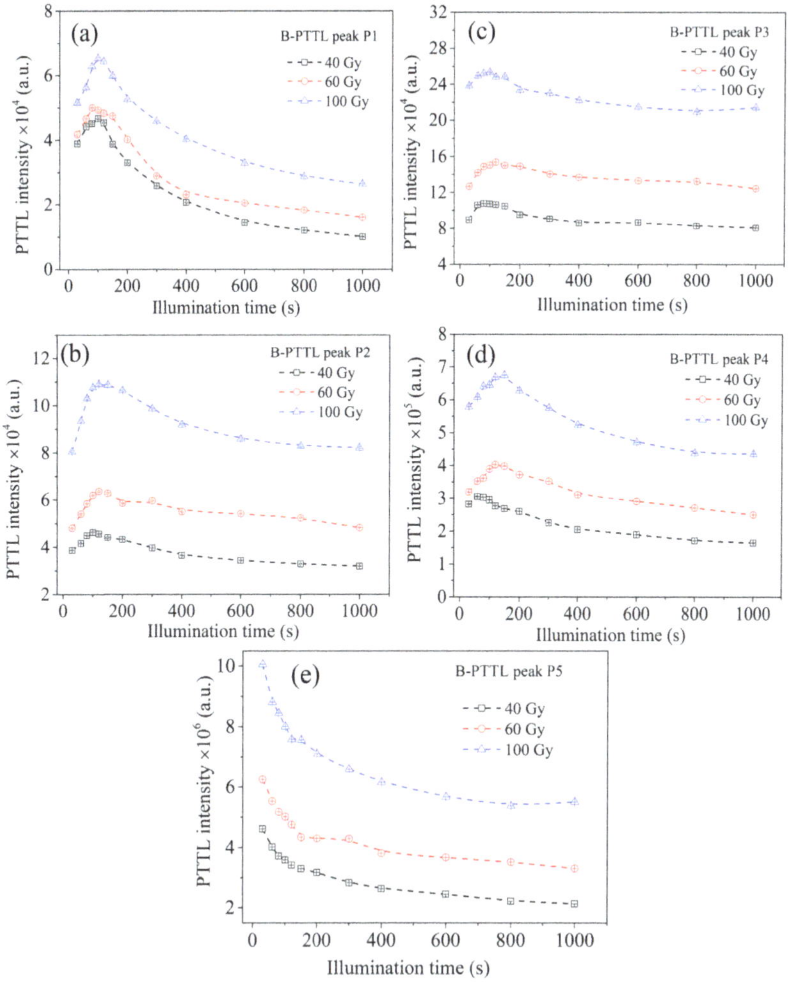

Figure 6.8. Change of PTTL intensity with blue light illumination for peaks P1 (a), P2 (b), P3 (c), P4 (d) and P5 (e) for the doses 40, 60 and 100 Gy. The dotted lines are visual guides only. Reprinted from [36], Copyright (2022), with permission from Elsevier.

For a general system of one acceptor and n donor electron traps, the transport of charge at the acceptor and at n donor electron traps each with electron concentration N_j, where $j = k;1\ldots, n$, can be expressed as given in equation (3.18). Using peak P1 as an example, its time-response profile was properly analysed as a system of two consequential donors, that is,

$$N_k = B_1(e^{-f_k t} - e^{-f_a t}) + B_2(e^{-f_k t} - e^{-f_b t}) \tag{6.1}$$

where $B_1 = (\alpha_a f_a N_{ai})/(f_a - f_k)$, $B_2 = (\alpha_b f_b N_{bi})/(f_b - f_k)$, where N_{ai} and N_{bi} are the initial concentration of electrons at the donor electron traps [36]. In dealing with a system of an indefinite number of donors as done here, the method described is necessarily an approximation since the number of donors n is unknown. In the method, the value of n is extended until a satisfactory fit is obtained. The minimum number of components that give an acceptable fit is used because increasing this further does not necessarily improve the standard deviation in the fitting statistics.

6.5.5 Dose response

The dose response of PTTL peaks P1, P2, P3 and P4 reproduced under blue light illumination is linear whereas that of peak P5 is sublinear for doses 10–100 Gy, preheating to 400 °C and illumination for 100 s [36]. This is shown in figure 6.9.

PTTL peaks obtained under infrared light stimulation are more intense than those following blue light illumination [36], as was also illustrated in figure 6.6. On

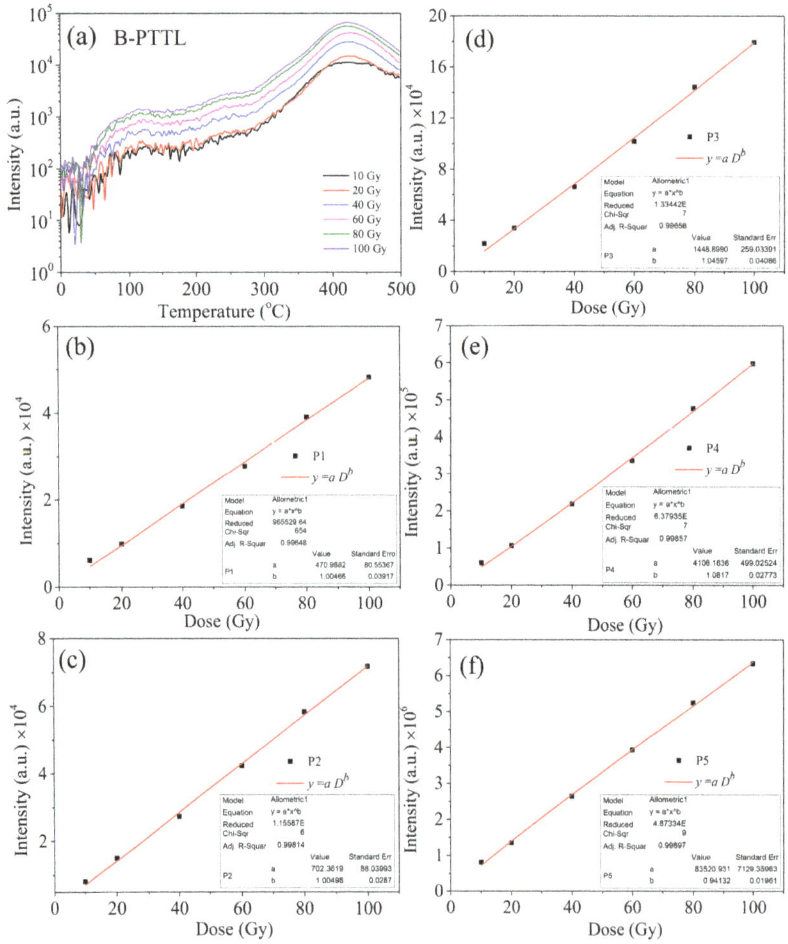

Figure 6.9. The dose response of peaks P1–P5 for the range 10–100 Gy following preheating to 400 °C and blue light illumination for 100 s. The line is in each case a fit of the function $y(D) = aD^b$ where a and b constants. Reprinted from [40], Copyright (2022), with permission from Elsevier.

the other hand, the dose response obtained following infrared light illumination is superlinear [36].

6.6 $SrAl_2O_4:Eu^{2+},Dy^{3+}$

6.6.1 Introduction

Strontium aluminate doped with Eu^{2+} and co-doped with Dy^{3+} ($SrAl_2O_4:Eu^{2+},Dy^{3+}$) is a widely studied inorganic phosphor (e.g. [1, 2, 38–40]). In comparison with conventional phosphors based on sulphides (e.g. ZnS), $SrAl_2O_4:Eu^{2+}Dy^{3+}$ produces longer lasting photoluminescence, which explains its enduring use in a range of eclectic applications including signage.

Studies on $SrAl_2O_4:Eu^{2+},Dy^{3+}$ although substantial, have been dominated by its synthesis and investigations on processes responsible for the spontaneously emitted luminescence such as photoluminescence and fluorescence. Exceptions are fewer and include investigations of its TL (e.g. [1, 2, 41, 42]), optically stimulated luminescence and the PTTL [39], and optical-ionization based TL [43].

The literature on $SrAl_2O_4:Eu^{2+},Dy^{3+}$ has long been replete with off-hand manipulation of TL data. Despite a number of reports (e.g. [2, 43–45]) showing that a simple peak in persistent luminescence phosphors including $SrAl_2O_4:Eu^{2+},Dy^{3+}$ may only be apparently so, the number of examples where the TL from $SrAl_2O_4:Eu^{2+},Dy^{3+}$ is analysed and described with indifference as to whether a particular methodology is valid is concerning. This could be due to the fact that now classic methods for experimentally separating peaks do not readily apply in this class of materials.

In a follow-up work to the study of Matsuzawa *et al* [38], Yamamoto and Matsuzawa [46] were categorical in stating that the TL of $SrAl_2O_4:Eu^{2+},Dy^{3+}$ consists of one dominant peak. The claim seems to have set a perspective that guided many subsequent studies on the TL of $SrAl_2O_4:Eu^{2+},Dy^{3+}$. Thus, the TL has sometimes been reported, analysed and described on the basis of techniques that apply to a single isolated peak (e.g. [47]). Indeed, the geometrical symmetry of the peak has also in some cases been used to speciously deduce that the peak follows second order kinetics.

The TL glow curve of $SrAl_2O_4:Eu^{2+}, Dy^{3+}$ (e.g. [1, 38, 39, 41]) consists of glow peaks with a wide expanse close to or exceeding, in some cases, 100°. The kinetic analyses previously done on such peaks imply a tacit assumption that the only component of the glow curve is the most prominent glow peak. However, some experimental studies of phosphors with similar features such as $Ca_5(PO_4)_3OH:Gd^{3+},Pr^{3+}$ [45] or $Sr_2SiO_4:Eu^{2+}$ [2, 42, 44] show a collocation of peaks. This is a situation where a dominant peak subsumes subsidiary ones to such an extent that the collection appears as one. Therefore, simply associating a prominent peak with a single discrete electron trap potentially disregards other peaks concealed within the peak profile and leads to incomplete conclusions about the mechanisms in the TL. It has been unequivocally demonstrated that the prominent peak in $SrAl_2O_4:Eu^{2+},Dy^{3+}$ and the glow curve consist of closely-spaced components (e.g. [1, 2, 42]). In terms of emission bands, measurements using x-ray excited optical luminescence show that stimulated luminescence from $SrAl_2O_4:Eu^{2+},Dy^{3+}$ has two prominent emission bands, one at 475 nm and a more intense one near 575 nm with weaker intensity emissions at 405, 510, 600 and 660 nm [42].

6.6.2 PTTL of $SrAl_2O_4:Eu^{2+},Dy^{3+}$

The PTTL of $SrAl_2O_4:Eu^{2+},Dy^{3+}$ related to blue light and infrared light illumination, respectively, was studied by Chernov [39, 48]. Supplementary measurements used infrared light. In the measurements, which used 470 nm and infrared light illumination [39] to induce phototransfer, the PTTL was detected in the 300–600 nm band [39]. Their glow curve showed four peaks at 340, 430, 560 and 680 K. It was noticed that once the 340 K and 430 K peaks were removed by preheating to 433 K and 533 K, exposure of the sample to infrared light partly restored the 340 K and 430 K peaks. The source traps were attributed to those corresponding to electron traps of the 430 K and 680 K peaks. The nature of the electron trap was discussed as relating to a Dy^{3+} ion at different Sr sites.

The role of light-induced charge transfer between electron traps in $SrAl_2O_4:Eu^{2+}$, Dy^{3+} was also reported by Van der Heggen *et al* [49]. Measurements were made on unirradiated material using 470 nm blue light with detection in the green part of the spectrum. Their glow peaks typify ones observed in $SrAl_2O_4:Eu^{2+},Dy^{3+}$ in being characteristically wide in expanse with two peaks standing out at 110 °C and 375 °C for heating at 1 °C s^{-1}. The protocol followed in this study was to expose the sample to repetitive 1 s light exposures after which the sample was heated to 250 °C to record the first apparent peak at 110 °C. After the set of n such exposures, the sample was then heated to 500 °C to record the second apparent peak. It was observed that the intensity of the 110 °C peak increased at the expense of the higher temperature peak. These results were only seen when the sample was illuminated with no changes for measurements that omitted illumination. The authors ascribed this to 'optical redistribution of trap occupation'. The results observed here are akin to phototransfer.

6.7 Selected applications

6.7.1 Gorilla glass

Samples of Gorilla® glass (Corning) from different types of touchscreen cellphone were analysed by McKeever *et al* [50] using PTTL in order to retrospectively assess radiation doses absorbed by the glass for potential applications in emergency dosimetry. PTTL was induced by (UV) light (365 nm). It was noted that the PTTL was stable over several days. The dose response curves (PTTL intensity as a function of initial applied dose) were linear up to about 20 Gy. A follow up study by Chandler *et al* [51] addressed various issues including the matter of background signals including a means to separate radiation-induced signals from background signals. The use of PTTL for emergency dose assessment was reviewed by Wrzesień *et al* [52].

6.7.2 Ge-doped SiO_2 optical fibre

The possibility of using PTTL from Ge-doped SiO_2 optical fibre for potential applications including dose reassessment in radiation dosimetry and in dating was reported by Zulkepely *et al* [53]. Phototransfer was achieved using a 254 nm UV source.

6.7.3 LiF-TLD 100

Bhasin *et al* [54] reported PTTL measured from LiF-TLD 100 induced using UV light. The phototransfer was attributed to a particular peak as a source. The significance of this work is that it also reported some issues that could not be easily explained by the simple model of PTTL and should thus serve as an impetus for further study.

6.7.4 Alumina substrates

The possibility of using alumina substrates, such as those utilised as surface-mount resistors in cellphones as dosemeter materials was investigated by Bossin *et al* [55]. PTTL generated using 470 nm light was used to sample deep donor electron traps and a detection limit of ca 100 mGy was found. The dose response was supralinear below 10 Gy. Using an alternative 307–575 nm broadband light source, a linear dose response was also found in the same dose range. Interestingly, the detection limit was higher at ca 200 mGy. This was attributed in part to a signal originating from the presence of a non-radiation-induced photostimulated TL signal.

6.8 Summary

This chapter dwelt on materials with glow curves made up of a quasi-continuous distribution of glow peaks. We dealt with three types of glow curves: those with collocated peaks, others with closely overlapping peaks and a third type that cannot be described as either of the first two. The purpose of the discussion was to illustrate how different experimentalists addressed the measurement and/or analysis of PTTL in such materials. We also picked out a few applications of interest. These are not meant to be exhaustive as there are many other interesting examples that had to be omitted for lack of space.

References

[1] Chithambo M L, Wako A H and Finch A A 2017 Thermoluminescence of $SrAl_2O_4:Eu^{2+}$, Dy^{3+}: kinetic analysis of a composite-peak *Radiat. Meas.* **97** 1–13

[2] Van den Eeckhout K, Smet P F and Poelman D 2010 Revealing trap depth distributions in persistent phosphors *Phys. Rev.* B **87** 045126

[3] Kalita J M and Chithambo M L 2020 Structural, compositional and thermoluminescence properties of microcline ($KAlSi_3O_8$) *J. Lumin.* **224** 117320

[4] Polymeris G S, Pagonis V and Kitis G 2017 Thermoluminescence glow curves in preheated feldspar samples: an interpretation based on random defect distributions *Radiat. Meas.* **97** 20–7

[5] Sfampa I K, Polymeris G S, Pagonis V, Theodosoglou E, Tsirliganis N C and Kitis G 2015 Correlation of basic TL, OSL and IRSL properties of ten K-feldspar samples of various origins *Nucl. Instrum. Methods Phys. Res.* B **359** 89–98

[6] Jain M, Guralnik B and Andersen M T 2012 Stimulated luminescence emission from localized recombination in randomly distributed defects *J. Phys. Condens. Matter* **24** 385402

[7] Rivera Montalvo T, Olvera Tenorio L, Azorín Nieto J, Barrera Salgado M, Soto Estrada A M and Furetta C 2005 Thermoluminescence characteristics of hydrogenated amorphous zirconia *Radiat. Eff. Def. Sol.* **160** 181–6
[8] Chithambo M L 2012 Dosimetric features and kinetic analysis of thermoluminescence from ultra-high molecular weight polyethylene *J. Phys. D: Appl. Phys.* **45** 345301
[9] Gribble C D 1988 *Rutley's Elements of Mineralogy* (London: Unwin Hyman)
[10] Gaft M, Reisfeld R and Panczer G 2005 *Luminescence Spectroscopy of Minerals and Materials* (Berlin: Springer)
[11] Pagonis V, Chen R, Kulp C and Kitis G 2017 An overview of recent developments in luminescence models with a focus on localized transitions *Radiat. Meas.* **106** 3–12
[12] de Farias Soares A, Tatumi S H, Courrol L C, de Faria Barbosa R and Kuruduganahalli Ramachandraiah N 2022 Studies on luminescence properties and photo-thermo transfer phenomena in fluorapatite *Radiat. Phys. Chem.* **201** 110490
[13] Fattahi M and Stokes S 2003 Dating volcanic and related sediments by luminescence methods: a review *Earth Sci. Rev.* **62** 229–64
[14] Rendell H M, Steer D C, Townsend P D and Wintle A G 1982 The emission spectra of TL from obsidian *Nucl. Tracks and Radiat. Meas.* **10** 591–600
[15] Alzahrani J S, Soliman C, Aloraini D A and Alzahrany A A 2016 Phototransferred thermoluminescence from obsidian using ultraviolet radiation *J. Nat. Sci. Res.* **6** 53–9
[16] McKeever S W S, Moscovitch M and Townsend P D 1995 *Thermoluminescence Dosimetry Materials* (Kent: Nuclear Technology Publishing)
[17] Forner L A, Viccari C and Nicolucci P 2020 Dosimetric properties of thermoluminescent pellets of $CaSO_4$ doped with rare earths at low doses *Radiat. Phys. Chem.* **171** 108704
[18] Guckan V, Ozdemir A, Altunal V, Yegingil I and Yegingil Z 2019 Studies of blue light induced phototransferred thermoluminescence in CaSO4:Mg *Nucl. Inst. Methods* B **448** 31–8
[19] Chithambo M L, Niyonzima P and Kalita J M 2018 Phototransferred thermoluminescence of synthetic quartz: analysis of illumination-time response curves *J. Lumin.* **198** 46–154
[20] Stoddard A E 1960 Effects of illumination upon sodium chloride thermoluminescence *Phys. Rev.* **120** 114
[21] Mckeever S W S 1985 *Thermoluminescence of Solids* (Cambridge: Cambridge University Press)
[22] Pedroza-Montero M, Melendrez R, Chernov V, Barboza-Flores M and Castaneda B 2002 Study of the phototransferred thermoluminescence in KCl:Eu^{2+} phosphors *Radiat. Prot. Dosim.* **100** 183–5
[23] Barboza-Flores M, Meléndrez R, Chernov V, Bernal R, Piters T M, R P-S, Aceves R, Pedroza-Montero M and Castaneda B 2001 Phototransferred thermoluminescence of KCL: Eu^{2+} dosemeters *APPC* **2000** 638–40
[24] Anthony J W, Bideaux R A, Bladh K W and Nichols M C (ed) 2003 *Handbook of Mineralogy* (Chantilly, VA: Mineralogical Society of America) pp 20151–1110
[25] Strickertsson K 1985 The thermoluminescence of potassium feldspars glow curve characteristics and initial rise measurements *Nucl. Tracks* **10** 613–7
[26] Visocekas R, Tale V, Zink A, Spooner N A and Tale I 1996 Trap spectroscopy and TSL in feldspars *Radiat. Prot. Dosim.* **66** 391–4
[27] Chruścińska A 2001 The fractional glow technique as a tool of investigation of TL bleaching efficiency in K-feldspar *Geochronometria* **20** 21–30
[28] Pagonis V, Morthekai P and Kitis G 2014 Kinetic analysis of thermoluminescence glow curves in feldspar: evidence of a continuous distribution of energies *Geochronometria* **41** 168–77

[29] Polymeris G S, Theodosoglou E, Kitis G, Tsirliganis N C, Koroneos A and Paraskevopoulos K M 2013 Preliminary results on structural state characterization of K-feldspars by using thermoluminescence *Mediter. Archael. Archaeom.* **13** 155–61

[30] Tachiya M and Mozumder A 1974 Decay of trapped electrons by tunnelling to scavenger molecules in low-temperature glasses *Chem. Phys. Lett.* **28** 87–9

[31] Tachiya M and Mozumder A 1975 Kinetics of geminate-ion recombination by electron tunnelling *Chem. Phys. Lett.* **34** 77–9

[32] Brown N D and Rhodes E J 2017 Thermoluminescence measurements of trap depth in alkali feldspars extracted from bedrock samples *Radiat. Meas.* **96** 53–61

[33] Kalita J M and Chithambo M L 2021 Blue-and infrared-light stimulated luminescence of microcline and the effect of optical bleaching on its thermoluminescence *J. Lumin.* **229** 117712

[34] Botter-Jensen L, McKeever S W S and Wintle A G 2003 *Optically Stimulated Luminescence Dosimetry* (Amsterdam: Elsevier)

[35] Robertson G B, Prescott J R and Hutton J T 1993 Bleaching of the thermoluminescence of feldspars by selected wavelengths present in sunlight *Nucl. Tracks Radiat. Meas.* **21** 245–51

[36] Kalita J M and Chithambo M L 2022 Phototransferred thermoluminescence characteristics of microcline ($KAlSi_3O_8$) under 470 nm blue- and 870 nm infrared-light illumination *Appl. Radiat. Isotopes* **181** 110070

[37] Chithambo M L 2021 Phototransferred thermoluminescence of tanzanite: a matrix-based analysis of time-response profiles and competition effects *J. Lumin.* **234** 117969

[38] Matsuzawa T, Aoki Y, Takeuchi N and Murayama Y 1996 A new long phosphorescent phosphor with high brightness, $SrAl_2O_4$:Eu^{2+},Dy^{3+} *J. Electrochem. Soc.* **143** 2670–3

[39] Chernov V, Mélendrez R, Pedroza M M, Yen W M and Barboza-Flores M 2008 The behaviour of thermally and optically stimulated luminescence of $SrAl_2O_4$:Eu^{2+},Dy^{3+} long persistent phosphor after blue light illumination *Radiat. Meas.* **43** 241–4

[40] Ntwaeaborwa O M, Nsimama P D, Shreyas P, Nagpure I M, Vinay K, Coetsee E, Terblans J J, Swart H C and Sechogel P T 2010 Photoluminescence properties of $SrAl_2O_4$:Eu^{2+},Dy^{3+} thin phosphor films grown by pulsed laser deposition *J. Vac. Sci. Technol.* A **28** 901–5

[41] Bedyal A K, Vinay K, Lochab S P, Singh F, Ntwaeaborwa O M and Swart H C 2013 Thermoluminescence response of gamma irradiated $SrAl_2O_4$:Eu^{2+}/Dy^{3+} nanophosphor *Int. J. Mod. Phys.* **22** 365–73

[42] Chithambo M L 2017 Thermoluminescence of the main peak in $SrAl_2O_4$:Eu^{2+}, Dy^{3+}: spectral and kinetics features of secondary emission detected in the ultra-violet region *Radiat. Meas.* **96** 29–41

[43] Bos A J J, van Duijvenvoorde R M, van der Kolk E, Drozdowski W and Dorenbos P 2011 Thermoluminescence excitation spectroscopy: a versatile technique to study persistent luminescence phosphors *Radiat. Meas.* **131** 1465–71

[44] Chithambo M L 2014 A method for kinetic analysis and study of thermal quenching in thermoluminescence based on use of the area under an isothermal decay-curve *J. Lumin.* **151** 235–43

[45] Mokoena P P, Chithambo M L, Vinay K H C, Swart H C and Ntwaeaborwa O M 2015 Thermoluminescence of calcium phosphate co-doped with gadolinium and praseodymium *Radiat. Meas.* **77** 26–33

[46] Yamamoto H and Matsuzawa T 1997 Mechanism of long phosphorescence of $SrAl_2O_4$:Eu^{2+}, Dy^{3+} and CaAl2O4:Eu^{2+}, Nd^{3+} *J. Lumin.* **72–74** 287–9

[47] Kshatri D S, Khare A and Jha P 2013 Thermoluminescence studies of SrAl2O4: Eu2 phosphors at different Dy concentrations *Chalc. Lett.* **10** 121–9

[48] Chernov V, Yen W M, Agundez-Arvizu Z, Melendrez R and Barboza-Flores M 2005 TL, IRSL and phototransferred TL in beta-irradiated $SrAl_2O_4$: Eu^{2+},Dy^{3+} ed K Worhoff, D Misra, P Mascher, K Sundaram, W M Yen and J Capobianco ed *Science and Technology of Dielectrics in Emerging Fields and Persistent Phosphors (PV 2005-13), Part II* (The Electrochemical Society) pp 231–48

[49] Van der Heggen D, Vandenberghe D, Moayed N K, De Grave J, Smet P F and Joos J J 2020 The almost hidden role of deep traps when measuring afterglow and thermoluminescence of persistent phosphors *J. Lumin.* **226** 117496

[50] McKeever S W S, Minniti R and Sholom S 2017 Phototransferred thermoluminescence (PTTL) dosimetry using gorilla glass from mobile phones *Radiat. Meas.* **106** 423–30

[51] Chandler J R, Sholom S, McKeever S W S and Hall H L 2019 Thermoluminescence and phototransferred thermoluminescence dosimetry on mobile phone protective touchscreen glass *J. Appl. Phys.* **126** 074901

[52] Wrzesień M, Al-Hameed H, Albiniak Ł, Maciocha-Stąporek J and Biegała M 2020 The photo-transferred thermoluminescence phenomenon in case of emergency dose assessment *Radiat. Environ. Biophys.* **59** 331–6

[53] Zulkepely N N, Amin Y M, Md Nor R, Bradley D A, Maah M J, Mat Nawi S N and Wahib N F 2015 Preliminary results on the photo-transferred thermoluminescence from Ge-doped SiO_2 optical fiber *Radiat. Phys. Chem.* **117** 108–11

[54] Bhasin B D, Kathuria S P and Moharil S V 1988 Some peculiarities of photo-transfer thermoluminescence in LiF-TLD 100 *Phys. Stat. Solid* A **106** 271–6

[55] Bossin L, Bailiff I K and Terry I 2018 Phototransferred TL properties of alumina substrates *Radiat. Meas.* **120** 41–6

www.ingramcontent.com/pod-product-compliance
Ingram Content Group UK Ltd.
Pitfield, Milton Keynes, MK11 3LW, UK
UKHW051853100926
8556IPUK00005B/602

9 780750 338295